FUKUSHIMA: DISPOSSESSION OR DENUCLEARIZATION?

FUKUSHIMA: DISPOSSESSION OR DENUCLEARIZATION?

Edited by

Majia Nadesan, Antony Boys, Andrew McKillop and Richard Wilcox

The Dispossession Publishing Group
2014

First published in 2014

By The Dispossession Publishing Group
http://wp.me/17PGi

First Printing: 2014

The preferred citation for this work is:

Nadesan, M., Boys, A., McKillop, A. & Wilcox, R. (Eds.) (2014), *Fukushima: Dispossession or Denuclearization?* The Dispossession Publishing Group

ISBN 978-1-312-49817-4

Cover artwork: WAR AND PEACE © WilliamBanzai7/Colonel Flick

* The Dispossession Publishing Group consists of the Editors of and Contributors to this book

Proceeds from sales of this book will be donated to

THE FUKUSHIMA COLLECTIVE EVACUATION TRIAL TEAM

http://fukushima-evacuation-e.blogspot.jp/ (English)

http://www.fukushima-sokai.net/ (Japanese)

http://fukusima-sokai.blogspot.jp/ (Japanese)

CONTENTS

Preface

POLITICS AND DISPOSSESSION

The dark shadow of nuclear power clouds the global energy, economic and national security outlook in 30 nations, and many more when the "upstream" of uranium mining, processing and transport is included. Due to costs, but also due to a previous ironclad political commitment to nuclear power, linked to the twin role of civil nuclear power and nuclear weapons production, we have a nuclear legacy, which in Japan has become a massive handicap. The nuclear threat to the economy, public health and the environment is an acute daily source of concern for Japanese, but the dispossessing role of nuclear power is massive – and complex.

Contributors to this anthology consider the dispossession threat of nuclear power. At its simplest, the "nuclear power system" is so convoluted, complex and costly, but so shrouded in secrecy, that abandoning it constantly throws up challenges. Some are expected and some not. Perhaps one of the greatest challenges concerns the secrecy surrounding the scope and severity of radiation's effects on human health. Clouded by censorship and deception, nuclear externalities are trivialized, manipulated, and outright denied, despite growing epidemiological and laboratory research demonstrating genetic and epigenetic vulnerability to ionizing radiation. In effect, nuclear dispossesses citizens of fundamental rights of personhood, including health.

Expensive – and Dangerous

Nuclear power started with unfounded and unrealistic hopes of "energy too cheap to meter," ignoring its extreme radiation dangers and basic role in producing nuclear weapons, but in reality is everywhere dependent on massive hidden or open subsidies from governments. These have reached extremes in many of the older nuclear countries now facing the daunting economic challenge of decommissioning and dismantling their aging, and increasingly unsafe, reactor fleets. Massive investments must also be made in long-term or "perpetual storage" centers for high-level wastes, which will remain lethal for centuries, or thousands of years ahead.

As contributors argue, the private corporate sector is both unable and unwilling to take on these challenges. It pursues a growth-or-bust strategy while its economic and financial performance continuously declines, making the nuclear power legacy a daunting handicap for the economy going forward through a looming perspective of a collapsed asset bubble comparable with the US subprime crisis of 2008. The nuclear subsidies so necessary for the industry's existence are fundamentally market distorting and disrupting.

Energy Transition – Forced or Voluntary

Moving to a high role for the renewable sources of energy and systematically increasing the energy efficiency of the economy are rational alternative to nuclear power. But contributors argue the ongoing nuclear crisis situation in Japan makes it difficult to separate hope, fears and exaggerated claims for the feasibility and practicality of energy transition going forward. Japan's energy transition has been forced – and jumpstarted – by the nuclear crisis. And it has not been fully embraced by a Japanese government still committed to nuclear power.

Our contributors look at how transition can be planned and managed in other countries, as well as Japan. The ongoing role of overpriced oil is considered by several contributors. Oil energy, like nuclear power is a legacy energy issue creating its own set of complex and difficult issues made worse by oil's role as a key financial asset in global asset trading. Leading countries for a non-nuclear future include Germany, with its vaunted Energiewende energy transformation plan aiming for the complete abandonment of nuclear power by January 2022. But Germany now faces serious and increasing technical, financial, economic, political and social challenges to Energiewende. Our contributors examine the German role and model in order to consider alternatives to nuclear dispossession.

Chapters

Fukushima: Dispossession or Denuclearization? aims to critique and transcend the nuclear energy paradigm. Contributors include individuals from diverse backgrounds and geographic locations, united by their common concern with the externalities produced by the nuclear industry. The project emerged out of electronic conversations held in the wake of the Fukushima disaster about the fundamental irrationality (i.e., madness) of nuclear energy. The conversations prompted the collection, with the intent to make a widely available e-book presentation of the fundamental economic, ecological, medical and social risks associated with nuclear energy. Discussion is far-reaching, but each chapter addresses the systemic risks posed by nuclear energy to a sustainable future, with special emphasis on the lessons learned from the Fukushima Daiichi disaster.

The book is divided into three distinct sections: *Politics, Dispossession, and Energy Transitions.* Politics explores the idea and practices of "nuclear dispossession," which is visually illustrated with the project's cover-art, created by William Banzai. Majia Nadesan's introductory chapter "From Hiroshima to Fukushima" explains the integral political-ideological relationship between atomic war and atomic energy visualized in Banzai's stunning "Hiroshima Fukushima" composition. Adam Broinowski's "Sovereign Power Ambitions and the Realities of the Fukushima Nuclear Disaster" argues that the inseparability of military and commercial nuclear programs in Japan's long-harboured desires for great power status is undermining the Nuclear Non-Proliferation regime and stifling public knowledge concerning the health effects from the Fukushima nuclear disaster. Richard Wilcox and Tony Boys conclude this section by investigating possibilities for, and challenges to, change with their chapter "The Political Challenge of Denuclearizing Japan."

Section II of the book – Dispossession – begins with Chris Busby's chapter on the health effects of nuclear fallout follows. Busby suggests that denial is built into commonly used dose-effects models used for predicting excess deaths and diseases caused by radiation exposure. Majia Nadesan's chapter "Fukushima and Dispossession: The End of Liberal Democracy in Japan?" considers the risks posed by nuclear to liberal democratic rights. Paul Langley concludes this section by describing how nuclear energy leads to the dispossession of memory itself as traumatic nuclear events are erased from public memory, even as they are re-enacted, as illustrated by the case of radioactive "black rain" found in Japan after the World War II Hiroshima and Nagasaki bombings and again after the March 2011 Fukushima Daiichi nuclear disaster.

Section III of the book – *Energy Transitions* – grounds the impetus for alternative energy with Harvey Wasserman's "50 Reasons We Should Fear the Worst from Fukushima." Next, Andrew McKillop's "European Energy Transition - Japan's Non-Nuclear Future" explains how Japan can simultaneously free itself of energy insecurity and nuclear power. Christian T. Lystbaek provides a business rationale for transition in "What is the Business of Business? CSR, the Fukushima Crisis and Energy Transition in a Changing World." Lystbaek argues that traditional models of corporate social responsibility (CSR) are inadequate for representing health and environmental externalities of business operations. He calls for a gestalt-switch in risk-assessment, with implications for the nuclear-energy paradigm. Social responsibility is really about human sustainability. Unfortunately, short-term profitability most often trumps all other organizational decisional criteria so the shift towards renewables may depend upon the efforts of private individuals. Tony Boys and Richard Wilcox draw upon interviews with Japanese residents to understand how citizens enact grassroots resistance against the nuclear-power paradigm through household decision making in "Grassroots Denuclearization: Can Japan Denuclearize by Adopting a Renewable Energy Future?" Their interviews focusing on rooftop photovoltaic (PV) panel reveal dissatisfaction with nuclear power and openness to renewables. They point out that Japan currently (as of June 2014) enjoys an opportunity for an energy-paradigm shift in the hiatus of the widespread nuclear shutdown produced by the Fukushima Daiichi disaster.

Contributors

As noted above, the contributors to this collection hail from diverse backgrounds. Despite differences in geography and training, contributors share a common commitment to a sustainable future based on renewable energy sources.

Moreover, each offers specialized insight into the Fukushima disaster based on their professional and/or scholarly expertise and personal experience.

WilliamBanzai7 is an artist/polemicist who specializes in creative satire and visual parody targeting all things concerning the financial industrial complex and related politics. He is a former professional and is knowledgeable with respect to most matters concerning regulatory capture and Wall Street chicanery. His works are published wildly at diverse internet venues, trading screens, cubby walls, refrigerators and survivalist bunkers throughout parts unknown. He calls his work: Visual Combat Banzai7. WilliamBanzai7 is of partial Japanese descent and has lived, studied and worked in Japan.WilliamBanzai7 prefers to publish under his 'nom de plume' as a professional courtesy to his former colleagues and clients. All of his work, spanning the period from the financial meltdown of September 2008 to date, can be found at his blog: www.williambanzai7.blogspot.com and on Flickr: www.flickr.com/photos/expd/.

Adam Broinowski is a postdoctoral research fellow at the School of Culture, History and Language at the Centre for Asia and the Pacific at the Australian National University. He holds a PhD from the Centre for Ideas and the School of Philosophical and Historical Studies, University of Melbourne. He has been a research fellow at the University of Tokyo, and lecturer at the Asia Institute and the VCA, University of Melbourne. His monograph is "Cultural responses to Occupation in Japan: The Performing Body during and after the Cold War" (Bloomsbury 2014). His Australian Research Council research fellowship at the ANU is "Contaminated Life: 'Hibakusha' in the Nuclear Age".

Tony Boys is British but has lived in Japan for nearly 40 years. He has an MA in International Studies from Tsukuba University and now works as a freelance translator in the countryside of Ibaraki Prefecture. He has been involved in alternative energy research for many years, particularly in the field of the connections between food production and fossil fuels in Japan.

Christopher Busby is an expert on the health effects of ionizing radiation. He qualified in Chemical Physics at the Universities of London and Kent, and worked on the molecular physical chemistry of living cells for the Wellcome Foundation. Professor Busby is the Scientific Secretary of the European Committee on Radiation Risk based in Brussels and has edited many of its publications since its founding in 1998. He has held a number of honorary University positions, including Visiting Professor in the Faculty of Health of the University of Ulster and was until recently Guest Researcher at Jacobs University Bremen, Germany. In his book Wings of Death (1995) he argued that the radiation risk models employed by national governments were unsafe for internal radionuclide exposures like those from Uranium weapons and Strontium-90, and he showed that the global cancer epidemic which began in 1980 was a consequence of the atmospheric nuclear tests of the 1960s. He followed this in 2007 with Wolves of Water, which examined cancer and radiological pollution of the Irish Sea, work funded by the Irish State. He has made several epidemiological studies of radiation effects, most recently in Fallujah, Iraq. Busby currently lives in Riga, Latvia. See also www.chrisbusbyexposed.org, www.greenaudit.org and www.llrc.org.

Christian T. Lystbaek is Associate Professor of Leadership and Organization Development at School of Business and Social Science, Aarhus University, Denmark. His primary research topics of interest are management development, corporate social responsibility, and business ethics. Christian's career spans two decades of working with organizations to transform their culture and processes away from command and control toward more reflective and collaborative work systems. Prior to joining Aarhus University he has worked as a leadership and organizational development consultant in a wide array of corporate environments including large and small business organizations and government agencies.

Andrew McKillop is an economist, research consultant and writer on environmental and energy issues. Among other notable posts, he has served as a senior research associate for the Science Council of Canada, National Energy Coordinator for the Government of Papua New Guinea, and Expert-Policy and Programmes, Energy Directorate, European Commission, Brussels. Andrew is author of *The Final Energy Crisis* with Pluto Press. He was first energy editor of the journal *The Ecologist* and has published works with other analysts, e.g. 'Oil Crisis and Economic Adjustment', Pinter Publishing, with Dr Salah al-Shaikhly, currently the Interim Iraqi government's Ambassador to London. His prolific essays on energy and the environment can be found widely, including on sites such as *Financial Sense* (http://www.financialsense.com/contributors/andrew-mckillop), among others.

Majia Holmer Nadesan is Professor of Communication Studies at Arizona State University's New College. Her scholarship in the areas of risk, biopolitics, political economy, and autism has been published in a wide variety of professional and peer-reviewed venues, including four academic books: *Fukushima and the Privatization of Risk* (Palgrave, 2013), *Governing Childhood: Biopolitical Strategies of Risk Management and Education* (Palgrave, 2010), *Governmentality, Biopower and Everyday Life* (Routledge, 2008), and *Constructing Autism* (Routledge, 2005). Musings about nuclear, financial, and environmental dispossession found on her blog, http://majiasblog.blogspot.com/.

Paul Langley is an independent scholar who has researched the public health effects of radiation for decades. Langley's research is based on close readings of official historical documents – including US and Australian government research reports – and detailed analysis of interviews conducted with atomic survivors. Langley authored (2012) *Medicine and the Bomb: Deceptions from Trinity to Maralinga* (available from: http://www.lulu.com/shop/paul-langley/medicine-and-the-bomb-deceptions-from-trinity-to-maralinga/ebook/product-21762622.html) and also regularly publishes on his nuclear history blog: http://nuclearexhaust.wordpress.com/.

Harvey Wasserman is author or co-author of a dozen books and edits the www.nukefree.org website. His Green Power & Wellness Show is at www.progressiveradionetwork.com. In 1973-4, he helped found America's grassroots "No Nukes" movement, a phrase he helped coin. He is senior advisor to Greenpeace USA and the Nuclear Information & Resource Service and Senior Editor of www.freepress.org. He speaks regularly to citizen and campus groups around the US. In 1994 he addressed 350,000 semi-conscious rock fans at Woodstock 2. Harvey teaches history at two colleges in central Ohio.

Richard Wilcox holds a Ph.D. in Environmental Studies from a social science, holistic perspective and teaches English at a number of universities in the Tokyo, Japan area. His articles on environmental topics including the Fukushima nuclear disaster are archived at *Reporting from Tokyo*, http://wilcoxrb99.wordpress.com/.

The Fukushima Five are a group of lifelong environmentalists and anti-nuclear advocates who published their first article on the Fukushima nuclear disaster on March 13th, 2011. It was entitled *Japan: A Nation Consigned To Nuclear Armageddon* and took the form of "An Open Letter to the People of Japan". Many subsequent articles and essays were written on the same subject, the most widely read was *As Fukushima Goes, So Goes Japan*. We share a strongly held conviction: The current *Nuclear Energy Paradigm* is fatally flawed, and therefore the world must transition away from it as a major source of energy.

POLITICS

Atomic Beginnings: From Hiroshima to Fukushima

Majia Holmer Nadesan

The alchemists' dream of elemental transmutation was achieved in the twentieth century through the deliberate and sustained deconstruction of atoms. Long slave to matter, man was made supreme, bending matter to his will, through the process of engineering atomic fission. Man had achieved the power of gods, or so it seemed. The Atomic Age was an age of hubris and, simultaneously, great existential unease as society registered the indisputable fact that the tool of ascendancy was simultaneously a tool of annihilation.

Fission and fusion and the unmapped interiorities of imploding/exploding matter were manufactured under Staff Field Stadium in Chicago on December 2, 1942. The earth – fertile and fierce – was soon humbled by the instrumentalization of atomic destruction with the Trinity explosion in New Mexico on July 16, 1945. Men worshipped their tools of creation and destruction. They strove to build ever more and more powerful instruments of annihilation. Earth scorching and DNA code-breaking radiation released in atomic chain reactions accelerated aging and destroyed fecundity among those downwind of the blasts triggered to affirm man's omniscience. But, evidence of destructive power was a terrible aphrodisiac that simultaneously prevented satiation. Man was enslaved by a terrible desire for (the power of) annihilation.

World War II was a spectacle of this enslavement. Men in many countries strategized to demonstrate omniscience through the atomic vaporization of their enemies. The Americans won the race by dropping two atomic bombs on Japanese cities. On August 6, 1945, the "Little Boy" atomic bomb was dropped on Hiroshima. The "Fat Man" bomb, dropped over Nagasaki on August 9, followed. The Americans fully understood the potential consequences for civilians and had actually investigated the weapons' potential for atomic fallout.

The fascination with destruction is illustrated five years later by a 1950 article published in *The Science News-Letter* describing the atomic bomb as a "Mass Murder Design." The article explained that the complete fission of only 2.2 pounds of uranium in an atomic bomb is equivalent to 20,000 tons of TNT. Gamma rays and neutrons from the bomb are described as a "wave of invisible energy" producing radiation sickness by "striking the single human cell in the bone marrow, the blood and the living tissues." Neutrons are described as "lethal up to a half a mile," while anyone within 3000 feet of the burst has only a 50 percent chance of surviving the radiation exposure," even if shielded by 12 inches of concrete (Matthews, 1950, pp. 122-123).

The world quickly understood the dangers of atomic explosions. Many nations were appalled by US President Dwight D. Eisenhower's readiness to use atomic weapons against civilian populations. But these very same nations were quickly developing their own atomic reactors and weapons programs because they too wanted the power of gods. The Atomic Age was inflected with a destructive mania, evidenced by the quantity and scale of bombs deployed in the atmosphere and earth. The destructive mania was calculating because it sought to insulate itself from mounting unease among the world's inhabitants regarding atomic contamination. Numerous historians have documented a deliberate strategy to propagandize on atomic matters. In particular, the propagandists sought to reassure civilians that atomic scientists' precise control over, and understanding of, radioactive fallout eliminated any risks to life or property. Popular atomic icons and technical, jargon-laden dose-effect models soothed the existential anxiety brought on by the spectacle of atomic annihilation (see Mackedon, 2010). Yet, unease prevailed.

In retrospect it seems remarkable that atomic energy could be launched in a context of atomic anxiety. However, atomic energy was specifically marketed to be the salve for public anxiety about atomic weapons. Stefan Possony, Defense Department consultant to the Psychological Strategy Board, had advised Eisenhower that "the atomic bomb will be accepted far more readily if at the same time atomic energy is being used for constructive ends" (Osgood, 2008, p. 156). Osgood documents the strategic push for atomic energy and the pivotal selection of Japan to illustrate atomic transmutation: Japan was transformed from atomic victim to atomic victor as the atomic reign of death was transformed symbolically into the atomic provider of life energy. Atomic Energy Commission Thomas Murray's comment illustrates the logic of replacement: "Now, while the memory of Hiroshima and Nagasaki remain so vivid, construction of such a power plant in a country like Japan would be a dramatic and Christian gesture which could lift all of us far above the recollection of the carnage of those cities" (cited in Tanaka & Kuznick, 2011).

The transformation was facilitated by Eisenhower's "Atoms for Peace" speech, delivered to the 470th Plenary Meeting of the United Nations General Assembly on December 8, 1953. The speech rhetorically transformed the horrors of atomic weapons into the productive, peaceful promise of atomic energy:

> The United States would seek more than the mere reduction or elimination of atomic materials for military purposes. It is not enough to take this weapon out of the hands of the soldiers. It must be put into the hands of those who will know how to strip its military casing and adapt it to the arts of peace.
>
> The United States knows that if the fearful trend of atomic military build-up can be reversed, this greatest of destructive forces can be developed into a great boon, for the benefit of all mankind. The United States knows that peaceful power from atomic energy is no dream of the future. The capability, already proved, is here today. Who can doubt that, if the entire body of the world's scientists and engineers had adequate amounts of

fissionable material with which to test and develop their ideas, this capability would rapidly be transformed into universal, efficient and economic usage?

Eisenhower outlined the US's leadership role in this transformation: "The United States would be more than willing -- it would be proud to take up with others 'principally involved' the development of plans whereby such peaceful use of atomic energy would be expedited." He tasked the United Nations with the creation of an international atomic energy agency responsible for monitoring and governing a stockpile of uranium and fissionable materials that could be employed for the "peaceful" development of atomic energy. Atomic energy was being heralded as provider of perpetual peace by eliminating energy resource scarcity (US NRC, 2012).

Atomic Promises Produce Atomic Perils

Eisenhower's speech is widely regarded in retrospect as a symbolic turning point; it rhetorically transformed the perils of the atomic bomb into the promise of a conflict-free atomic energy era. However, atomic horrors were not so easily swept away. In 1954, fallout from a US test explosion at Bikini went astray, prompting evacuation of Marshall Islanders from their homes (Walker, 2000). Health effects were indisputable and included "whitened" hair, nausea, diarrhea, itching and burning skin, watery eyes, hair loss, widespread skin lesions, and blood changes that lasted as long as six months. The propaganda machine was active. A *The New York Times* article reassured, "Fall-Out Effects Gone in 6 Months: 5 Navy Doctors Tell A.M.A: Pacific Blast Caused Mainly Skin Damage:" "Persons accidently showered with radioactive fallout in the nuclear detonation in the Pacific on March 1, 1954 recovered in six months from their major ailment—skin damage" (Plumb, 1955, p. 21). Despite this "mishap," the United Nations upheld the US "right" to conduct hydrogen bomb testing in the Pacific. The petition brought forward by the Marshall Islanders that testing be stopped was disregarded. The US Atomic Energy Commission (AEC), tasked with oversight of atomic issues, argued that the tests were critical for US security (Walker, 2000).

In addition to demonstrating Northern disdain for indigenous southern hemisphere inhabitants, the Marshall Islands accident illustrated limits to man's actual control over atomic fission. Atomic scientists and engineers could not predict with certainty fallout volume, nor migration. Fallout – tiny particles of matter connected to radioactive elements – migrates according to the whims of nature, evading experimental control. Scientific authorities were also limited in their understanding of immediate and long-term fallout effects for animals, plants, and humans. The US AEC's insistence that all forms of radiation exposure be represented publicly in sunshine units further confounded scientific and lay appreciation of how radionuclide effects derive from their concentration up the food chain (Miller, 1986, p. 203). The Marshall Islanders discovered the limits of scientific understanding of fallout effects: "jelly fish babies" were born without bones to islanders inhabiting areas deemed safe by US atomic authorities (Pacific Island Report, 2005).

Control over atomic decay was fundamentally limited. Yet, the seduction of power implied in harnessing atomic energy was too great to resist. Scientific man was made a demigod by his capacity to deconstruct atoms for war and peace. The public was encouraged to worship the atom and its keepers in popular culture. In Japan, Osamu Tezuka produced a manga series from 1951 to 1968 titled, *Tetsuwan Atom*, or "Mighty Atom," featuring a heroic robotic boy fueled by fission. An animated film series ran from 1963-1966. Yomoto Inuhiko explains that the main character, Atom Boy (also known as Astro Boy), was transformed over time from a shunned hero who yearned to be human (like Pinocchio), into "a faultless champion of justice." The narrative tone, especially in the television broadcasts, ideologically promoted scientific progress (cited in Utsumi, 2012, pp. 188-189). Atomic science was the new religion. But it was not uniformly embraced. The public was eager to adopt atomic medicine and yearned after the promise of endless energy, but was simultaneously haunted by the devastation caused by uncontrolled fission. Popular culture reflected this tension. In the US, the children's periodical *Adventures in Science* disclosed nuclear perils and promises through the story of a boy who loses his dog in an atomic test site ("Come Worship Atomic Energy," 2012). The narration matches the implied horrors of an irradiated dog with the promise of atomic medicine. In the end, the narrative reassures readers that their fears of radioactive fallout are overhyped as the irradiated dog is returned to its youthful owner after a mere one-week quarantine. Despite such carefully couched assurance strategies, the public remained conflicted and uncertain about atomic beneficence. Having seen the power of an atomic explosion, many among them were troubled by promises of atomic "safety."

The public had been educated by newspapers, popular periodicals, and book-length manuscripts to understand that the same "chain reaction" process was at the heart of both the atomic bomb blast and atomic energy production: the chain reactions of atomic fission were integrally explosive, whether employed for war or peace. For example, David Dietz's accessible 1954 *Atomic Science, Bombs and Power* explained on page 3 that "Like the atomic bomb, the reactor depends upon the creation of a chain reaction in uranium or plutonium." According to Dietz, science editor of Scripps Howard newspapers, the difference being that the latter is "controlled," whereas the former is not. The US boiling water reactors being developed at the time used water to slow the speed of neutrons, which mediated the critical chain. In contrast, the first "fast reactor" built at Los Alamos in 1946 was described by Dietz as "an atomic bomb under control" because it employed no moderator and relied on liquid mercury for cooling (p. 230). Thus, the chain process in an atomic plant required moderation to prevent an explosion.

Control over fission processes was also executed by scientists through the fuel ratio of Uranium 235 and Uranium 238 atoms. Uranium ore typically contains 0.7 percent of the fissile Uranium 235 atoms. Uranium fuel used in an atomic bomb would by design be "enriched" with more fissionable Uranium 235 than found in uranium fuel designed for civilian power production. Therefore, atomic scientists believed they could control the deconstruction of atoms through the choice of moderator and the degree of Uranium 235 enrichment. Man's atomic alchemy was sufficient to tame the demon of destruction that was fission.

Despite such careful education as provided by *Atomic Science, Bombs, and Power*, Atoms for Peace wasn't a successful whitewash. The problems of atomic accidents and waste remained stumbling blocks to atomic promises while amplifying risk of atomic weapons proliferation. Used fuel from civilian boiling water reactors can be mined for fissile elements, such as Uranium-235 and Plutonium-239. This process, described as "re-processing," was launched at Oak Ridge National Laboratory in the 1940s in order to isolate plutonium for atomic bomb production. Critics argued from the 1940s onward that reprocessing of fuel used in civilian reactors would enable atomic weapons proliferation. Indeed, in the September 11, 1948 issue of *The Science News-Letter,* Watson Davis argued that "every atomic power plant becomes potentially an atomic bomb material factory, from which there could be bootlegged the materials for illicit atomic bombs. Thus atomic power plants must be controlled if there is to be international or other control of atomic energy" (pp. 170-171). This concern about atomic weapons proliferation through purportedly "peaceful" atomic power remains today.

Weapons proliferation was not the only fear stirred by Atoms for Peace. Many critics were skeptical of atomic power because of basic engineering safety challenges. The same *Science News Letters* article warning of proliferation also outlined the significant safety challenges facing atomic power (Davis, 1948). The article emphasizes that plants designed for atomic energy production must be built of special materials capable of handling high heat levels, ranging upwards to 2,700 degrees Fahrenheit, and these special materials must also be capable of absorbing radiation, including neutron radiation:

> The deadly and intense radiation from nuclear fission must be protected against at all steps in atomic power production. This means thick shielding of concrete or other radiation-absorbing materials. The liquid picking up the heat in the atomic furnace will be almost as dangerous as the pile itself and the whole system must be leak-proof, which is much to ask because of the damaging effects of radiation upon machinery. Control rods in the pile (controlling the fission reaction) must be operated with great reliability inside the shielding and at the high temperatures. Replacements and repairs of the furnace will be dangerous because of the radiation contamination. Atomic power plants will be like battle-ships subjected to atomic bomb attack that become so "hot" they must be sunk at sea as a safety measure. (Davis, 1948, pp. 170-171)

The technical and proliferation challenges of atomic power were well understood. Yet, the seductions of atomic energy outplayed scientific caution and public anxiety. The perils were deemphasized, while the promise of limitless security was widely propagated by political rhetoric and industry salesmen.

Atomic Energy Scramble

By the late 1940s, the world's powers were scrambling to build atomic reactors for purposes both peaceful and warlike despite safety and security risks. Reactor designs differed across countries. The Canadian National Research Experimental Reactor, which became operational in 1947, used heavy water, containing the hydrogen isotope deuterium, as a moderator. In 1958 Canada instituted an atomic power division that produced the CANDU design (Canadian Deuterium Uranium) (Pringel & Spigelman, 1981). In 1956, Britain's Windscale air-cooled, plutonium-fueled, breeder reactor became operational. Britain declared its dual-purpose Windscale plant as the first to "produce electricity from atomic energy on a full industrial scale" (Aldred & Stoddard, 2008). Windscale's air-cooled design was atypical, as most reactors developed in Britain and France during the 1960s were gas-cooled. In *The Nuclear Barons*, Pringle and Spigelman (1981) argue that conflicts within Britain and France about the relative desirability of gas-cooled, versus light-water, reactor designs damaged their export industry, allowing the US boiling water reactor industry to dominate the international field.

Atomic cooperation grew throughout the 1950s. In 1952, ten western European nations formed the European Council for Nuclear Research. In 1954, the US amended its Atomic Energy Act to enable it to assist other nations in the development of their atomic power facilities and in 1955 it took the lead in drafting the Statute of the International Atomic Energy Agency (IAEA) with participation from government representatives from Australia, Belgium, Canada, France, Portugal, South Africa, the United Kingdom (IAEA, 1997). In 1956 the USSR, Czechoslovakia, India, and Brazil contributed representatives.

Development of atomic energy in the immediate post-war context did not always proceed smoothly and the media slowly began reporting on atomic mishaps. *The New York Times* reported that in 1952 an "atomic plant" at Chalk River Ontario had "caused peril" resulting in the "worst nuclear reactor accident that has been disclosed" (Plumb, 1954, p. 35). In July 1955, *The Manchester Guardian* observed that Dr. McCullough, Chairman of the Advisory Committee on Reactor Safeguards for the US AEC, warned publicly that reactors were far too dangerous to locate in populous areas. He was quoted in the article as stating:

> Many of us feel that this record [of few accidents] is just due to plain good luck, and our luck may not hold. We should be prepared for an accident. The thing we should try to avoid is a really bad accident. The key to the whole business is that reactors manufacture extremely poisonous materials, rather

worse by a million or billion times than anything else ever known. Though we have tried, we can find no valid comparison. It is a brand-new problem to us. (cited in Freedman, 1955, p. 9)

The belated nature of Dr. McCullough's admissions is noted in the text of the article, which reads: "There is something anomalous, he [McCullough] conceded, in worrying so much about safety regulations after the programme has been under way for some time." McCullough's greatest concern about reactor safety was continued heat creation, known as "delayed heat," that occurs after a reactor has been shut off. The safety challenge was to develop a reactor that could automatically shut off and control delayed heat production. McCullough warned that "we can't depend entirely on gadgets" to resolve this significant safety problem. His warnings were prescient. In 1957, the breeder reactor at Windscale in Britain caught fire and approximately 20,000 curies of radioactive iodine were acknowledged to have been released into the atmosphere (Weinberg, 1979). Details were suppressed for three decades, although Windscale was specifically designed to produce plutonium and likely released a wide array of radioisotopes during the fire.

The problem of reactor safety, and delayed heat specifically, would never be fully resolved, as the Fukushima crisis revealed. Fears about accidents and radiation releases were the flies in the atomic energy ointment despite media reassurances and the establishment of regulatory agencies. All types of atomic reactors can suffer from loss of cooling accidents and from "reactivity excursion," where control of the reactor core is lost, resulting in a runaway reaction (Caldicott, 2006, p. 124). Breeder reactors are particularly prone to these types of accidents because they operate at high temperatures requiring special mediators. Because of their high heat, breeder reactors produce more "fissile" elements (e.g., Plutonium-239) than produced in the fuel of the cooler-operating boiling water reactors (Makhijani & Saleska, 1991). Breeder reactors held the elusive promise of closing the fuel cycle: the fissile material produced by the breeder reactor could be re-used as fuel after reprocessing. However, no nation has successfully closed the fuel cycle because breeders proved to be highly explosive (von Hippel, 2010). Yet, many atomic scientists and policy leaders remained wedded to the breeder reactor project despite engineering and safety challenges.

Indeed, advocates of atomic energy have long viewed boiling water and other "thermal" reactors fueled with low enriched uranium as a "transitional step toward the use of breeder reactors" (Solomon, 1993, p. 1). Over 20 breeder reactors were built since 1951 in the US, France, UK, USSR, Japan and Germany. As noted above, France and England were especially committed to commercial breeder reactor programs in the early days of the scramble for atomic power until engineering challenges and accidents forced program closures. For example, France's sodium cooled and plutonium fueled Superphénix reactor, which operated between 1986 and 1997, was closed due to sodium leaks and cracks in the reactor vessel. It had been beset by problems throughout its tenure (Public Citizen, no date). The US suffered a meltdown in one of its fast breeder reactors, the Fermi 1 prototype fast breeder reactor, on October 5, 1966. The Soviet BN-350 fast breeder reactor suffered a sodium-water reaction accident in 1972. Although these countries still maintain some fast-breeders for military purposes, they have abandoned the promise of commercial breeder reactors.

Japan has for decades been especially emphatic in its support of commercial breeders and has pursued experimental breeder reactor programs and fuel reprocessing as part of its goal to create a closed-loop fuel cycle capable of meeting national energy needs (Sweet, 1988). Permission for Japan's Tokai reprocessing plant was granted in 1970 and the first reprocessing occurred in 1977 (Ichii, n.d.). Japan has accumulated vast stockpiles of plutonium, which is designated for experimental and future use in especially engineered fast breeder reactors (Ichii, n.d.). However, Japan has also experienced significant mechanical and safety challenges. Japan's $12 billion experimental "Monju" sodium-cooled, fast-breeder reactor in Fukui Prefecture was shut down soon after being activated by a major fire in 1995 (Daly, 2012). The operator attempted to hide the incident by having workers alter their report and through the creation of a strategically truncated video of the accident ("Monju Costs," 2012).

After public disclosure of the extent of problems with the breeder program, Japan Atomic Energy Commission concluded publicly in February 2012 that technological considerations prevented the civilian-breeder program from being a realistic option. The commission recommended that a more viable alternative would be to reprocess spent fuel for plutonium that could be used instead of Uranium-235 in the production of plutonium-uranium mixed oxide (MOX) reactor fuel for boiling water reactors. MOX burns hotter than ordinary uranium-based fuel and therefore poses additional safety considerations despite widespread promotion by the commercial atomic energy industry. Fukushima Daiichi Unit 3 reactor was running MOX fuel at the time of the March 2011 earthquake. Plutonium from the Fukushima Daiichi plant was detected in Lithuania in 2011 (Lujanienė, Byčenkienė, Povinec, & Gera, 2011).

Regulation and Safety in Atomic Industry

Atomic unease was not vanquished by the promises of endless energy and atomic medicine. Safety agencies were instituted and propaganda campaigns waged to promote public confidence in an inherently dangerous endeavor. In many countries the same agency was responsible for promoting *and* regulating atomic endeavors. For instance, the US established the AEC in 1946 and tasked it with all atomic matters (US NRC, 2012). Britain's Atomic Energy Authority (AEA), established in 1954, was responsible for ensuring the safety of atomic plants and for promoting atomic research. It too was institutionally positioned to promote atomic power while regulating it. In the wake of the Windscale accident, Britain established the Nuclear Installations Inspectorate (NII) as an independent government inspector in 1959. The US similarly created an independent agency, the Nuclear Regulatory Commission, in 1974. Agencies' role in promoting expansion of the atomic complex, while simultaneously guaranteeing plant safety, would come to be seen by many as a conflict of interests. Japan did not create a separate and independent agency to regulate atomic energy until 2012, with the September establishment of the Nuclear Regulation Authority.

Whether independent or not, most atomic agencies were and remain well-connected with other powerful government agencies. In the 1950s an atomic "priesthood" emerged composed of physicists and administrators in the growing and global industrial complex. This priesthood largely dictated policy for atomic developments (Grossman, 2012). Although some within the priesthood wanted atomic energy to remain a government monopoly due to safety concerns, others felt that the future lay with private commercialization (Lanouette, 1985). Government subsidization was the norm, even among countries that pursued commercial atomic energy. And the atomic powers in both industry and government sought also to export their tools as carrots for closely aligned subordinate nations. Illustrating the atomic sales pitch by government is President Eisenhower's negotiator on atomic matters, Morehead Patterson, who pledged the US would "do everything it can" to enable American businesses to build plants abroad ("Building A-Plants," 1955). The growing US atomic industry, led by General Electric and Westinghouse, helped promote atomic energy overseas with US intelligence initiatives and with the assistance of the IAEA. Indeed, nearly one third of all reactors sold by General Electric and Westinghouse between 1956 and 1974 were sold overseas (Hertsgaard, 1983). The first reactor exports were sold to Germany in 1958, Italy in 1959, and Japan in 1960. General Electric collaborated in licensing with Hitachi and Toshiba in Japan in 1967. The US Government Export-Import Bank facilitated the sale of atomic energy plants – primarily light-water designs – and fuel to countries oversees with $2.4 billion in loans and guarantees by 1974 (Hertsgaard, 1983).

The IAEA played an important role in promoting atomic energy for western countries and their industries. It set up a Department of Technical Assistance and a Joint Division with the U.N. Food and Agricultural Organization (FAO). The IAEA also inaugurated the Internal Centre for Theoretical Physics, which trained scientists from the developing countries and supported their research. It also organized and promoted international conferences on the "peaceful uses of nuclear energy," the first of which was held in August 1955. The IAEA was and continues to be a vigorous proponent of atomic energy. It also endows research at MIT through funded research grants, a practice that began in 1960 and continues today (IAEA, 1997). Grants helped support applied and theoretical research aimed at improving engineering safety and understanding the effects of ionizing radiation.

Despite expert assurances of safety and the obvious business opportunities, doubts prevailed because liability for accidents appears as an agenda item at the IAEA's February 1955 meeting (IAEA, 1997). Atomic plant safety and long-term storage of used fuel dominated the IAEA's public agenda. In May 1962 the IAEA held its first major symposium on reactor safety and in June of that year the agency approved its Basic Safety Standards for Radiation Protection.

Considerable media promotion of the emerging regulatory complex may not have deterred public fears, but it did help legitimize a growing industrial complex, allowing rapid expansion domestically within the US and overseas.

The World Nuclear Organization claims that the operational safety of the plants built in the 1970s was much improved over earlier reactors. Research on atomic engineering of plants across the 1970s through the 1990s period focused on radiation shielding, control systems, alarm systems, and hydrogen mitigation systems. Atomic plants in the 1970s had three physical barriers designed to prevent radiological contamination (Weinberg, 1979). First, metallic zirconium encased the fuel pellets. Second the actual core of the reactor containing the fuel rods, the zirconium encased fuel pellets, was crafted of thick steel designed to withstand significant pressure. Third, a concrete-encased steel dome served as a pressure-resistant outer containment vessel. These three levels of protection purportedly guaranteed against a core meltdown of the fuel rods.

The emphasis on safety was largely a result of ongoing atomic energy mishaps. In 1975, a fire disabled the emergency core cooling system at Browns Ferry in Alabama. In 1979, a meltdown at Three Mile Island resulted in evacuations and mass civil suits as a result of the release of radiation. The disaster at Three Mile Island marked a turning point in the US and elsewhere despite efforts to re-invigorate the nuclear industry. The realities of massive cost over-runs for utilities converged with strong anti-nuclear sentiments to erode utilities' enthusiasm for building more plants, particularly when they concluded coal was cheaper. Furthermore, pressure was mounting over how and where to store waste safely. Every year approximately one third of all rods in reactors are replaced (Lanouette, 1985). Removed "spent" rods are highly radioactive and must be cooled. "Spent fuel" rods with high radioactivity were accumulating in storage ponds at reactors, posing significant security risks (Alvarez et al., 2003). Concerns about thermal pollution of lakes and rivers from plants added to public distrust.

The 1986 Chernobyl plant disaster in the Ukraine cemented public resistance to atomic energy across many nations. The unprecedented scale of that disaster, its undeniable visibility, and the necessity for permanent evacuations of entire communities raised questions about catastrophic risks. Who assumes those risks in the event of disasters? Most countries with atomic energy have liability legislation that spells out and limits liabilities for accidents. The end result of liability limitations is that the externalities of atomic energy, particularly of accidents, are primarily born by impacted citizens (Nadesan, 2013). Public resistance to atomic energy has therefore been relatively strong, despite nations' efforts to promote it as vital for national security.

Japan's Will to Power

By the turn of the twentieth century, Japan had firmly adopted a will to modernize in order to avoid becoming subjugated by the western colonial powers expansion in Asia (Sapir & Van Hyning, 1956). Japan's hierarchical social structure facilitated the rapid, centralized push towards industrialization. By the time World War II began, Japan had a sophisticated scientific infrastructure that was employed to study atomic weapons. Under the leadership of Dr. Yoshio Nishina, the Japanese Army's Aviation Technology Research Institute employed a thermal diffusion process to separate Uranium-235 in 1943 (Dong-Joon & Gartzke, 2006). Imperial Japan understood well the power of the atom.

At the end of World War II, the occupying US banned all fission research. However, Japanese interests still coveted atomic power. In 1952, a newly created Japanese Science Council recommended that the Japanese government promote fundamental research in physics with the creation of a Research Institute of Nuclear Physics (Wit & Clubok, 1956). By 1953, the national Japanese budget included atomic energy appropriations for atomic research. The US clearly approved of these developments and substantiated that approval with the November 14, 1955 Agreement for Cooperation between Tokyo and Washington, which included US atomic assistance. That year, the Japan Atomic Energy Section was created within the Industrial Technology Board of the Ministry of International Trade and Industry. The next year, in 1955, an Atomic Energy Basic Law was passed and a Japanese Atomic Energy Commission was founded. A 1956 assessment of "Japanese Atomic Power Development" lists a vast array of studies underway involving heavy water exchange processes, fast neutron scintillators, and uranium extraction. In 1967, the Japanese Atomic Energy Commission's (JAEC) first "Long Term Plan" promoted fast breeder reactors as the mainstream for Japan's atomic energy future (Suzuki, 2010).

Atomic energy mediated US-Japan Post-World War II relations. The US gained an atomically capable Asian ally in its Cold War chess game. Indeed, in 1954, T. E. Murray of the US AEC told the steel union that atomic reactors in Japan were critical for the US to win the "Atom Race with Russia" ("T.E. Murray Nuclear Reactor Urged," 1954, p. 14). American manufacturing workers were likely receptive because Atoms for Peace created a ready market for US industry, including the domestic atomic giants, such as General Electric. Japan gained a fast-track to all the promises of atomic power, including purportedly endless energy and the international prestige and security of being perceived as atomic-weapons capable. As explained by Jane Nakano (2012), Japan's bilateral relationship with the US enabled Japan quick access to sophisticated technology, including breeder reactors:

> The 1955 bilateral agreement freed Japan from its earlier vision to incrementally and gradually acquire nuclear technologies, and generated the momentum for more rapid and wide-scale advancement of its nuclear energy program, including indigenous production of a Breeder Reactor in the longer term. (p. 44)

Japan's interest in breeder reactor programs ultimately became emblematic of the nuanced forms of security encoded in Japan's atomic research. Breeder reactors seduced LDP government strategists because of the promised energy independence and military might.

Despite the institutional "atomic" push, many Japanese citizens distrusted the entire atomic enterprise. Japan was heavily bombed during World War II, first with incendiary bombs and eventually with two atomic bombs. The high loss of civilian life was tolerated by the allies because the Japanese had been so dehumanized in war propaganda (Stone & Kuznick, 2012). Video footage of the carnage caused by the atomic bombs in Hiroshima and Fukushima was so horrifying that it was censored by the US military, as documented by Greg Mitchell's (2012) *Atomic Cover-Up*. Many among Japan's people adopted a strong anti-atomic attitude as a result of these horrific experiences, and by the accidental irradiation of Japanese fishermen in the Bikini Atoll in 1954. For example, the 1956 review of "Japanese Atomic Power Development" published in the journal *World Politics* notes that "many Japanese scientists, have argued that the United States is employing its Atoms for Peace program merely to screen from world opinion its military use of nuclear energy" and that the U.S objective was self-interested, aimed at promoting its industry and/or capturing control of foreign uranium resources (Wit & Clubok, p. 530).

The US responded to Japanese anti-nuclear sentiment with a massive propaganda campaign, which was launched in Japan on November 1, 1955 with a US exhibit heralding the peaceful applications of the atom. Matsutaro Shoriki, a founder of Japan's Liberal Democratic Party in 1955, worked with the CIA to successfully publicize the exhibit, and atomic power more generally, using his position as head of *The Yomiuri Shimbun* (Warnock, 2012). Shoriki is attributed with producing a shift in support for atomic energy using his strategic influence over Japanese print and television media. Indeed, *The Economist* refers to him as the Japan's Citizen Kane ("Japan's Citizen Kane," 2012). In 1954, Shoriki invited John Jay Hopkins of General Dynamics and Laurence Hafstad of Chase Manhattan to lecture on atomic energy and an "Atomic Marshall Plan" for Japan (Pringle & Spigelman, 1981, p. 173). Shoriki encouraged the import of atomic plant technology from the US and Britain. He dismissed concerns about safety from within the Japanese scientific community while pushing forward breeder technologies. Shoriki played a vital role in forging public acceptance for atomic energy by linking it to energy security publicly, while less publicly bringing Japan into the "nuclear club" of countries with access to enriched uranium and plutonium.

Shoriki's drive for atomic energy has been linked directly to his more militant nationalist pursuits. Shoriki was a right wing nationalist with militant tendencies ("Japan's Citizen Kane," 2012). He joined the Metropolitan Police in 1913, whereupon he gained notoriety for suppressing student demonstrations at Waseda University. After promotion to head of the Metropolitan Police, Shoriki infiltrated labor groups and Koreans who were agitating against Japanese colonization. He also may have incited mob violence against these groups, resulting in the violent death of Koreans living in Japan. His militancy and fascism resulted in imprisonment after World War II. Shoriki believed that Japan lost the war because of its technological inferiority. Atomic energy seemed to solve Japan's energy problems while affording it technological expertise and materials required for building atomic weapons. Shoriki's newspaper and Nippon television station (established 1952) were effective propaganda tools to promote atomic energy and Shoriki himself. Soon after his election to parliament, Shoriki established an Atomic Energy Authority, which later evolved into the Department of Science and Technology. By 1957, Japan had contracted to purchase 21 commercial reactors, most of which were of US design.

The character and historical development of Japan's electrical utilities facilitated rapid expansion of atomic energy within the nation. Japan's earliest utilities were privately held and regional. By the 1920s hydropower was the main source of energy for the regional utilities. A boom in utility producers in the 1920s resulted in steep competition and subsequent industry consolidation. By 1928, five companies had consolidated control of the industry: The Tokyo Electricity Company (TELC), the Daido Electricity Company, the Ujigawa Electricity Company, and the Nippon Electricity Company (Sapir & Hyning, 1956). During the 1920s through the end of World War II, Japan's utilities became integrally connected with wartime enterprises and were financially protected by the government. In 1939, the government consolidated remaining utilities by establishing the Japan Electric Power Generation and Transmission Company as a national monopoly. This powerful wartime monopoly was dissolved because of its concentrated economic power in 1951. In 1952 the Electric Power Development Promotion Law facilitated division of ten separate common electric power companies: Hokkaido Electric Power Company, Tohoku Electric Power Company, Tokyo Electric Power Company (TEPCO), Chubu Electric Power Company, Hokuriku Electric Power Company, Kansai Electric Power Company, Chugoku Electric Power Company, Shikoku Electric Power Company, Kyushu Electric Power Company, and Okinawa Electric Power Company (the last being added in 1972) (Navarro, 1996). These utilities became powerfully entrenched entities because each was a regional monopoly and they were all united as members of the Federation of Electric Power Companies.

All of these regional monopolies operate atomic energy plants, except for Okinawa Electric Power Company. In the early 1960s, Tokyo Electric Power Company and Kansai Electric Power Company, purchased American boiling water designs from General Electric and Westinghouse (Pringle & Spigelman, 1981). Japan's LDP government incentivized its regional utilities' forays into atomic energy by offering special depreciation allowances and tax cuts. TEPCO's Fukushima Daiichi Power Station was built in the 1960s by Ebasco, an American general contractor that no longer exists (Shirouzu & Dawson, 2011). Reactors 1 through 5 at the site were based on General Electric's Mark I design.

Two engineers from General Electric resigned in 1975 after concluding General Electric's Mark I reactors were flawed from the beginning and posed operational risks ("Fukushima: Mark One," 2011). Boiling water reactors operate with intense pressure and the engineers felt that the design specifications were insufficient for handling pressures that would result from a loss of cooling accident. One of the engineers, Dale G. Bridenbaugh, explained the engineers' concerns in an interview with ABC: "The problems we identified in 1975 were that, in doing the design of the containment, they did not take into account the dynamic loads that could be experienced with a loss of coolant . . . The impact loads the containment would receive by this very rapid release of energy could tear the containment apart and create an uncontrolled release." General Electric's poor reactor design likely contributed to the explosions that occurred at the Daiichi complex.

Atomic mishaps and concerns about safety grew in Japan well before the Daiichi accident. TEPCO has been particularly beset with scandals. In 2002, TEPCO's president, vice president, and chairman stepped down after the utility acknowledged that it failed to report accurately cracks at its reactors in the 1980s and 1990s ("Heavy Fallout," 2002). TEPCO was suspected of falsifying twenty-nine cases of safety repair records (Shirouzu & Tudor, 2011). The Japanese Nuclear and Industrial Safety Agency claimed that up to eight reactors could be operating with unfixed cracks, "though the cracks don't pose an immediate threat." TEPCO falsified coolant water temperatures at the Fukushima Daiichi plant in 1985 and 1988 ("Japan's Nuclear Power Operator," 2011). The falsified records were used during a 2005 inspection. In the wake of these scandals, TECPO revealed that an uncontrolled chain reaction had occurred in Unit 3 at Fukushima Daiichi when fuel rods fell into the reactor (Shirouzu & Smith, 2011). TEPCO has had problems at other sites as well. In July 2007, TEPCO's Kashiwazaki-Kariwa plant in Niigata Prefecture was damaged in a magnitude 6.8 earthquake ("New Japanese Nuclear Power," 2008). TEPCO claimed no radiation was released, but later admitted that radioactive water had spilled into the Sea of Japan (Shirouzu & Tudor, 2011).

General Electric design flaws, operator malfeasance, and Japan's geological activity all contributed to the Fukushima disaster. Unsafe conditions at atomic plants persisted across decades, as illustrated by TEPCO's problems at Daiichi, discussed above. Yet, throughout Japan, networks of powerful individuals and groups promoted a false "myth" of atomic safety, according to Harutoshi Funabashi (2012), a Japanese sociologist, who has elaborated upon how institutional collusion and media collaboration perpetuated complacency about reactor safety within Japan. Japan's utilities have regional monopolies and are backed by the institutional power of Japan's Ministry of Economy, Trade and Industry (METI). As explained previously, Japan's Nuclear and Industrial Safety Agency (NISA), which was located under METI authority, both regulated and promoted atomic energy until a new "independent" agency was created in 2012, the Nuclear Regulation Authority. Centralization of regulation and promotion within one agency undermined regulatory functions, a problem recognized publicly in the late 1990s with the scandals surrounding TEPCO. Adding to the electrical companies' power are networks of influence cultivated by industry grants and personal honorariums. Cultivated networks of influential actors help shape public opinion at national and regional levels. They have been active over the years in pushing municipalities to accept nuclear facilities. In a separate analysis of institutional collusion in Japan conducted prior to the Fukushima disaster, Valentine and Sovacool (2010) describe how institutional capture of the media occurs through mainstream journalists' press membership in "reporter's clubs." Membership in these clubs is implicitly conditional upon reporters' willingness to temper public criticism, particularly in the arena of atomic energy, as Valentine and Sovacool explain: "the combination of the LDP's majority lock on parliament and this unique form of media control served to keep the dangers of nuclear power development from the public eye" (p. 7973). Japan's atomic energy complex is clearly entrenched throughout Japanese industrial, research, and media institutions. It may also be linked to Japan's military security.

Japan's powerfully entrenched atomic energy industry and its extensive breeder reactor program signify internationally that Japan is an atomic-weapons capable nation. Japan has stockpiled plutonium for decades and reprocesses commercial fuel at Tokai (von Hippel & Takubo, 2012). Japan has been criticized for stockpiling more plutonium than it could use in a breeder reactor program or for mixed oxide fuel containing plutonium (MOX). As of 2010, Japan had more than 46 tons of separated plutonium stored domestically and in Europe (Suzuki, 2010). Japan's rationale for purchasing plutonium from the UK and France was never disclosed, beyond Japan's use of it in its civilian energy program (Takagi & Nishio, 1990). Japan's stockpile of plutonium would enable it to make five thousand warheads (Williamson, 2012), which could be delivered by Japanese rockets. In 2008 Japan passed the Aerospace Basic Law of 2008, which included a provision allowing development of space technology in relation to its "contribution to national security" ("Nuclear Law's," 2012). Japan's constitutional restriction on weapons development is being challenged openly by the LDP, particularly since their ascent back into power with the 2012 elections. On June 23, 2012, *The Mainichi* newspaper called for repeal of a national security clause that was embedded into Japan's amended Atomic Energy Basic Law, which passed on June 20, 2012 ("National Security Clause," 2012). *The Mainichi* described the clause as allowing "the possibility of nuclear armament open to interpretation," ("Atomic Energy Law," 2012). LDP's atomic commitments are also illustrated by its pledge to re-start Japan's commercial reactors that were shut down after March 11, 2011 ("Optimism Rises," 2012). The LDP remains committed to the controversial new Rokkasho reprocessing plant, despite geologists' concerns about a potentially active fault under the site (Hasegawa, 2012). An accident at the site could produce many fatalities (von Hippel & Takubo, 2012). These commitments indicate that the LDP worships a cult of death that is called "atomic energy" or "nuclear," in close collaboration with a similar American death cult.

Atomic Insanity

The pages of popular periodicals published during the atomic era demonstrate clearly that the atomic priesthood was well aware of the DNA code-destroying capacities of ionizing radiation. Yet, hubris reigned among the atomic death cult, as illustrated so poignantly by Robert Oppenheimer's famous interview about the Trinity explosion, broadcast in the 1965 documentary, "The Decision to Drop the Bomb":

> We knew the world would not be the same. A few people laughed, a few people cried, most people were silent. I remembered the line from the Hindu scripture, the Bhagavad-Gita. Vishnu is trying to persuade the Prince that he should do his duty and to impress him takes on his multi-armed form and says, "Now, I am become Death, the destroyer of worlds." I suppose we all thought that one way or another.

The push for atomic bombs and energy despite the known risks illustrate man's hubris and technocratic disregard for life. The atom is the apple of our fall: Man sought absolute power/knowledge in *his* control over the atom; he used this knowledge to fuel *his* insatiable will to power. In *his* hubris, man failed to acknowledge fully the terrible costs wrought by his atomic reign and now risks destroying the eco-system's carrying capacity for humanity, while simultaneously deconstructing human DNA itself. We surge toward apocalypse with disasters such as Fukushima, Chernobyl, and Hanford.

Atomic propaganda blinds us to our passage as we are soothed by the atomic priesthood's partial truths and misinformation, especially about the dangers of ionizing radiation and chemical toxicity. It is true that life evolved on earth because of radiation. Across time, radiation-induced mutations played an integral role in natural selection. However, what is conveniently forgotten is that any increase in exposure to radiation produces an increase in mutations and changes in cellular reproduction. The decay and chemical toxicity of radioisotopes, such as Uranium-235 and Strontium-90, embedded in bodily tissues pose special risks for health and reproduction. Caldicott (2011) explained in an editorial titled "Unsafe at Any Dose" that the idea of a threshold of no effects has been debunked scientifically and medically. Low doses can produce genetic damage and epigenetic changes. A significant increase in the average human's "body burden" of radionuclides and their decay products (e.g., lead) across a few generations could overwhelm evolutionary-derived repair mechanisms, producing chromosomal instability and genetic mosaicism (see Nadesan, 2013). Radiation-induced damage to reproductive DNA is transmitted across generations. Caldicott (2013) explains that radiation-induced mutations can take up to twenty generations to be fully expressed. The accumulation of recessive mutations across generations could trigger a fundamental state change in DNA integrity, resulting in cascading instabilities capable of precluding viable life. In 1956, geneticists involved in the US Biological Effects of Atomic Radiation (BEAR) analysis of atmospheric fallout effects warned of precisely this danger (US National Academy of Sciences, 1956; see Nadesan, 2013).

Many societies have explored the idea of a human-wrought apocalypse in their religions, mythology, and literature. The human capacity for collective madness is well-established in our apocalyptic tales. We recognize that the will to power is linked closely to a drive for human self-destruction. This truism has been documented especially in the will to atomic power. Those that organize the collective madness of atomic power (weapons and energy) employ propaganda to engineer popular consent around their death cult. The myth of atomic omniscience persists despite accident after accident in part because physicists and others among the priesthood use arcane language and symbolism to confuse public understanding of risk, especially by denying bio-accumulation, bio-magnification and transgenerational effects. The apocalypse may be near, but denial and self-destruction continue to reign.

References

Aldred, J., & Stoddard, K. (2008, May 26). Timeline: Nuclear power in the United Kingdom: key events in the history of nuclear power in Britain. *The Guardian.* Available http://www.guardian.co.uk/environment/2008/jan/10/nuclearpower.energy

Alvarez, R., Beyea, J., Janberg, K., Kang, J., Lyman, E., Macfarlane, A., Thompson, G. & F. von Hippel (2003). Reducing the hazards from stored spent power-reactor fuel in the United States. *Science and Global Security, 11*(1), 1–51.

Atomic Energy Law's sly alteration is abuse of legislative process. (2012, June 26). *The Mainichi.* Available http://mainichi.jp/english/english/perspectives/news/20120626p2a00m0na004000c.html

Building A-plants abroad gets green light. (5 April 1955). *The Washington Post and Times Herald.* Available http://login.ezproxy1.lib.asu.edu/login?url=http://search.proquest.com/docview/148680882?accountid=4485

Calidcott, H. (2013, November 1), Radiation fears are real. Available http://www.helencaldicott.com/2013/11/radiation-fears-are-real/

Caldicott, H. (2011, May 1). Unsafe at any dose. *The New York Times.* Available http://www.nytimes.com/2011/05/01/opinion/01caldicott.html?_r=0

Caldicott, H. (2006), *Nuclear power is not the answer.* New York: The New Press.

Chinnery, P. F., Samuels, D. C., Elson, J., & D. M. Turnbull (2002). Accumulation of mitochondrial DNA mutations in ageing, cancer, and mitochondrial disease: is there a common mechanism? *The Lancet,* 360(9342), 1323-5.

Come worship atomic energy at the church of 1950s science – classics illustrated special issue: Adventures in science (February 26, 2012). *Blog into mystery.* Available http://blogintomystery.com/2012/02/26/come-worship-atomic-energy-at-the-church-of-1950s-science-classics-illustrated-special-issue-adventures-in-science/

Daly, J. (27 February 2012). Another Fukushima causality: Japan's fast-breeder reactor program. *Oil Price.Com.* Available http://oilprice.com/alternative-energy/nuclear-power/another-fukushima-casualty-japans-fast-breeder-reactor-program.html

Davis, D. (1948, September 11). Atomic power production. *The Science News-Letter,* 54(11), 170-171.

Dietz, D. (1954). *Atomic science, bombs and power.* New York: Dood, Mead & Company.

Dong-Joon J. & Gartzke, E. (2006). *Codebook and data notes for determinants of nuclear weapons proliferation: A quantitative model.* Available http://pages.ucsd.edu/~egartzke/data/jo_gartzke_0207_codebk_0906.pdf

Drubek, K. (2012, July 26). In Japan, a nuclear ghost town stirs to life. *Open Channel NBC News.Com.* Available http://openchannel.nbcnews.com/_news/2012/07/26/12839675-in-japan-a-nuclear-ghost-town-stirs-to-life?lite

Eisenhower, D. (1953, December 8). Atoms for Peace. Address to the 470th Plenary Meeting of the United Nations General Assembly *International Atomic Energy Association.* Available http://www.iaea.org/About/history_speech.html

Freedman, M. (1955, July 15). Dangers of nuclear radiation: U.S. plans for public protection. *The Manchester Guardian.* Available http://search.proquest.com.ezproxy1.lib.asu.edu/docview/479866824/136B74FF0E260A4EEE8/3?accountid=4485

Fukushima: Mark 1 nuclear reactor design caused GE scientist to quit in protest. (2011, March 15). ABC the Blotter. Available http://abcnews.go.com/Blotter/fukushima-mark-nuclear-reactor-design-caused-ge-scientist/story?id=13141287

Funabashi, H. (2012) Why the Fukushima nuclear disaster is a man-made calamity. *The International Journal of Japanese Sociology, 21,* 65-75.

Grossman, K. (2012, June). Atomic orthodoxy and the ascent to heaven. *Enformable.* Available http://enformable.com/2012/06/atomic-orthodoxy-and-the-ascent-to-heaven/

Hasegawa, K. (2012, December 19). Quake risk at Japan atomic recycling plant. *Pys.Org.* Available http://phys.org/news/2012-12-quake-japan-atomic-recycling-experts.html#jCp

Heavy fallout from Japan nuclear scandal. (2002, September 2). *CNN.* Available http://archives.cnn.com/2002/BUSINESS/asia/09/02/japan.tepco/index.html

Hertsgaard, M. (1983) Nuclear Inc.: The men and money behind nuclear energy. New York: Pantheon.

Ichii, N. (n.d.). Japan experience on reprocessing in Japan by Japan's Agency for Natural Resources and Energy, Ministry of Economy, Trade and Industry. Available http://www.iaea.org/INPRO/2nd_Dialogue_Forum/Experience_on_Reprocessing_in_Japan.pdf

Inuhiko, Y. (1995). Hanseiki ni Wataru Jidai no Uneri to Kakuto Shita "Tezuka Manga" no Chiheri" [the Place of Tezuka Manga Struggling with Undulation of the Times for a Half Century]. In Tezuka Production & Kazuhiko Murakami (Ed.), *Tezuka Osamu ga Inakunatta Hi [The Day Tezuka Osamu Went Out]* (pp. 240-241). Tokyo: Shiode Supan.

International Atomic Energy Association (IAEA) (1997, September). IAEA turns 40: Supplement to the IAEA Bulletin. Available http://www.iaea.org/Publications/Magazines/Bulletin/Bull393/Chronology/chronology.pdf

Japan's Citizen Kane: A media mogul whose extraordinary life still shapes his country, for good and ill. (2012, December 22). *The Economist*. Available http://www.economist.com/news/christmas/21568589-media-mogul-whose-extraordinary-life-still-shapes-his-country-good-and-ill-japans

Japan's nuclear power operator has checkered past. (2011, March 12). *Reuters*. Available http://www.reuters.com/article/2011/03/12/us-japan-nuclear-operator-idUSTRE72B1B420110312

Kuznick, P. (2011, April 13). Japan's nuclear history in perspective: Eisenhower and atoms for war and peace. *The Bulletin of Atomic Scientists*. Available http://thebulletin.org/japans-nuclear-history-perspective-eisenhower-and-atoms-war-and-peace-0

Lanouette, W. (1985). Atomic energy: 1945-1985. *The Wilson Quarterly*, 9(5), 100-110.

Lujanienė, G., S. Byčenkienė, P.P. Povinec, M. Gera M. (2011, December 27). Radionuclides from the Fukushima accident in the air over Lithuania: Measurement and modeling approaches. *Journal of Environmental Radioactivity*, *114*, 71-80

Makhijani, A., & Saleska, S. (1999). *The Nuclear power deception*. New York: The Apex Press.

Matthews, S. (1950, August 19). Mass murder design. *The Science News-Letter*, *58*(8), 122-123.

Miller, R. L. (1986). *Under the cloud: The decades of nuclear testing*. New York: The Free Press.

Mitchell, G. (2012). Atomic cover-up: Two US soldiers, Hiroshima & Nagasaki and the greatest movie never made. New York: Sinclair Books.

Monju costs far surpass usual nukes: Trouble-prone reactor has rung up far higher tab than initially planned. (2012, July 4). *Japan Times*. Available http://www.japantimes.co.jp/text/nn20120704f1.html

Nadesan, M. (2013). *Fukushima and the privatization of risk*. Houndmills, UK: Palgrave Pivot.

Nakano, J. (2012). Civilian nuclear energy cooperation between the United States and Japan. In Y. Tatsumi (Ed.), *The new nuclear agenda: Prospects for US-Japan cooperation*. Available http://www.stimson.org/images/uploads/research-pdfs/New_Nuclear_Agenda_FINAL_3_15_12.pdf

National security clause must be deleted from law on atomic energy. (2012, June 23). *The Mainichi*. Available http://mainichi.jp/english/english/perspectives/news/20120623p2a00m0na009000c.html

Navarro, P. (1996). The Japanese electric utility industry. In R. J. Gilbert & E. P. Kahn (Eds.), *International comparisons of electrical regulation* (pp. 231-276). Cambridge: University of Cambridge.

New Japanese nuclear power reactors delayed. (2008, March 26). *World Nuclear News*. Available http://www.world-nuclear-news.org/NN-New_Japanese_nuclear_power_reactors_delayed-260308.html

Nuclear law's "national security" clause must be dropped. (2012, June 22). *The Asahi Shimbun*. Available http://ajw.asahi.com/article/views/editorial/AJ201206220037

Optimism rises in "nuclear village" after LDP's victory. (2012, December 19). *The Asahi Shimbun*. Available http://ajw.asahi.com/article/0311disaster/fukushima/AJ201212190048

Oppenheimer, J. R. (1965). Interview. In F. Freed (Producer), *The decision to drop the bomb*. NBC White Paper. Available Atomic Archive, http://www.atomicarchive.com/Movies/Movie8.shtml

Osgood, K. (2008). Total Cold War. Eisenhower's secret propaganda battle at home and abroad. Lawrence, KS: University of Kansas Press.

Pacific Island Report. (2005, June). *Nuclear nightmare lingers in Marshalls*. Available http://archives.pireport.org/archive/2005/June/06-14-com2.htm

Plumb, R. K. (1955, June 9). Fall-out effects gone in 6 months. *The New York Times*, p. 21.

Plumb, R. K. (1954, December 1). Reactor accident caused peril in 1952: Flooding of atomic plant with deadly radioactive water in Canada is disclosed here. *The New York Times*, p. 35.

Pringle, P., & Spigelman, J. (1981). *The nuclear barons* (2nd ed). New York: Avon.

Public Citizen (no date). *Fast reactors: Unsafe, uneconomical and unable to resolve the problems of nuclear power*. Available https://www.fas.org/programs/ssp/_docs/FastReactors.pdf

Sapir, M. & Van Hyning, S. J. (1956). *The productive uses of nuclear energy: The outlook for nuclear power in Japan*. Washington, DC: National Planning Association.

Shirouzu, N., & Tudor, A. (2011, March 15). Crisis revives doubts on regulation. *The Wall Street Journal*. Available http://online.wsj.com/article/SB100014240527487033639045762005337461 95522.html

Shirouzu, N., & Dawson, C. (2011, July 1). Design flaw fueled nuclear disaster. *The Wall Street Journal*, pp. A1, A12.

Shirouzu, N., & Smith, R. (2011, March 16). Plant's design, safety record are under scrutiny. *The Wall Street Journal*. Available http://online.wsj.com/article/SB100014240527487043965045762044619299 92144.html

Solomon, K. A. (1993). *Plutonium for Japan's nuclear reactors: Paying both for the proliferation and dollar price to assure long-term fuel supply*. Santa Monica, CA: RAND/UCLA Center for Soviet Studies.

Sugimoto, T. (2012, July 24). After 500 days, Fukushima no. 1 plant still not out of woods. *The Asahi Shimbun*. Available http://ajw.asahi.com/article/0311disaster/fukushima/AJ201207240087

Suzuki, T. (2010, February). Japan's plutonium breeder reactor and its fuel cycle. In T. Cochran, H. Feiveson, W. Patterson, G. Pshakin, M. Ramana, M. Schneider, T. Suzuki, & F. von Hippel (Eds.), *Fast breeder reactor programs: History and status: A research report of the International Panel on Fissile Materials* (pp. 53-61). Available http://fissilematerials.org/library/rr08.pdf

Sweet, W. (1988). Japan's nuclear program stresses breeders, plutonium, and safeguards. *Physics Today*, *41*(1), 71-74.

Takagi, J., & Nishio, B. (1990). Japan's fake plutonium shortage. *The Bulletin of Atomic Scientists*, *46*(8), 34-38.

Tanaka, Y. & Kuznick, P. (2011, May 2). Japan, the atomic bomb, and the "peaceful uses of nuclear Power." The Asia-Pacific Journal, 9(18). Available *http://japanfocus.org/-Yuki-TANAKA/3521*

Taylor, R. W. & Turnbull, D. M. (2005). Mitochondrial DNA mutations in human disease. *Nature Reviews Genetics*, *6*(5), 389–402. Available http://www.ncbi.nlm.nih.gov/pmc/articles/PMC1762815/

T. E. Murray nuclear reactor urged for Japan: T. E. Murray of A. E. C. tells steel union step is vital in atom race with Russia. (1954, September 22). *The New York Times*, p. 14.

U.S. National Academy of Sciences (1956*). Committees on* Biological *Effects of Atomic Radiation, 1954-1964.* Available http://www.nasonline.org/about-nas/history/archives/collections/cbear-1954-1964.html

U.S. Nuclear Regulatory Commission (2012, March 19). History. *USNRC.* Available http://www.nrc.gov/about-nrc/history.html

Utsumi, H. (2012). Nuclear power plants in "the only bombed country": Images of nuclear power and the nation's changing self-portrait in postwar Japan. In D. van Lente (Ed.), *The nuclear age in popular media: A transnational history, 1945-1965* (pp. 175-202). Basingstoke, UK: Palgrave.

Valentine, S. & Sovacool, B. (2010). The socio-political economy of nuclear power development in Japan and South Korea. *Energy Politics, 38,* 7973-7979.

von Hippel, F. (2010, February). Overview: The rise and fall of plutonium breeder reactors. In T. B. Cochran, H. A. Feiveson, W. Patterson, G. Pshakin, M.V. Ramana, M. Schneider, T. Suzuki, & F. von Hippel (Eds.), *Fast breeder reactor programs: History and status: A research report of the International Panel on Fissile Materials* (pp. 1-16). Available http://fissilematerials.org/library/rr08.pdf

Von Hippel, F., & Takubo, M. (2012, November 28). Japan's nuclear mistake. *The New York Times.* Available http://www.nytimes.com/2012/11/29/opinion/japans-nuclear-mistake.html?nl=todaysheadlines&emc=edit_th_20121129&_r=0

Walker, S. (2000). *Permissible dose: A history of radiation protection in the twentieth century.* Berkeley, CA: University of California Press.

Warnock, E. (2012, June 1). Japan's nuclear industry: The CIA link. *The Wall Street Journal.* Available http://blogs.wsj.com/japanrealtime/2012/06/01/japans-nuclear-industry-the-cia-link/

Weinberg, A. M. (1979). Nuclear energy: Salvaging the Atomic Age. *The Wilson Quarterly, 3*(3), 88-93.

Williamson, P. (2012, May 31). Plutonium and Japan's nuclear waste problem: International scientists call for an end to plutonium reprocessing and closing the Rokkasho plant. *The Asia Pacific Journal.* Available http://japanfocus.org/-Piers-_Williamson/3766

Wit, D. & Clubok, A. B. (1956). The United States and Japanese atomic power development. *World Politics, 8*(4), 515-533.

Sovereign Power Ambitions and the Realities of the Fukushima Nuclear Disaster

Adam Broinowski

In seeking to comprehend why the Abe administration has not done more to stem the nuclear disaster at the Fukushima Daiichi nuclear power plant, it has been necessary to identify the relationship of the disaster with a major and ongoing shift in the status of sovereign power relations, both in Japan and elsewhere. With the ongoing nuclear disaster having served as a reminder of the systematic influence of the transnational 'nuclear village', the radical politics of the Abe government are commensurate with a broader shift in Japan's strategic posture to meet perceived geopolitical dynamics in the Asia-Pacific region. In the nexus of corporate investment, nuclear power plants, resource distribution, military technology and geopolitical relations, it is instructive to observe how Japan's evolving combined military and nuclear power policies, from the 1950s and intensifying in the 1990s in particular, underpin the official denial of public health dangers posed by ongoing leakage of radioactive contaminants onto land and into the water and food supply, and into the ocean. It is my contention here that unless state-corporate policies which prioritize geopolitical ambitions and economic interests can be held accountable to the rights of civil society and international law, such policies will continue to be based on criteria which fundamentally destabilize the essential structures which comprise planetary life.

Extended Sovereign Power

The disaster at the Fukushima Daiichi nuclear power plant on 11 March 2011 occurred at a time when it was increasingly obvious that the constitution of 'sovereign power' was changing. In a democratic system, sovereign power has been commonly understood as that which is vested in a government elected by the people to enact decisions in their fundamental interests and to protect their well-being through its delegated authorities. As the ongoing disaster at Fukushima Daiichi has shown, however, it seems that sovereign power in advanced capitalist Japan, is also delegated to a range of institutions and executive authorities which are not accountable to the public and which do not hold its interests as their main priority.

Japan's centralised energy structure which is monopolised by ten electricity utilities and in which nuclear power is a main pillar, suggests a different understanding of 'the sovereign'. In the continuing nuclear crisis caused by the meltdowns/melt-throughs/blow-outs of four General Electric nuclear reactors owned by Tokyo Electric Power Company (TEPCO), it is clear that a broader 'nuclear village' (*genshiryoku mura*) of an influential closed-circle of policy makers including regulatory agencies and government, electricity monopolies, component and plant manufacturers and pro-nuclear knowledge institutes and media outlets are the constituents of an extended notion of sovereign power (Hara in Hindmarsh, 2013, 22-40; Samuels 1987, 237-8). The concerted effort to protect and consolidate the 'village' after 3.11 demonstrates the extent of its influence. government and broadcast networks, which receive revenue from these electricity monopolies, as do state institutions and corporations beyond Japan's national borders, have sought to secure their interests in the beleaguered TEPCO. In fact, as the Abe government has pursued an agenda of domestic nuclear re-starts and continued with planned nuclear projects, it has been under the influence of the US Department of Energy (DoE) and the Nuclear Suppliers Group (NSG) comprising of 48 nations to promote Japanese nuclear technologies overseas. This suggests how a transnational nuclear village influences the decisions of a sovereign state as an extended form of sovereign power.

The Hobbesian conception of sovereign power is understood as the right vested by its subjects in the sovereign as the interpreter of God's will to 'kill and let live' (Foucault 1990, 135-6, 178). In Giorgio Agamben's reading of Carl Schmitt's *Political Theology* (1922) and Walter Benjamin's *Critique of Violence* (1921) (Benjamin in Bullock and Jennings, 1996; Schmitt 2006), this legal right is carried out either under a 'state of exception' in which special laws permit sovereign intervention, or a legal suspension which allows for extra-judicial intervention in exceptional circumstances (Agamben 2005, 52-64). This necessarily includes the capacity to determine for the national interest a sacrificial population during a state of emergency.

As a divine monarch no longer rules Japan, the elected government must exercise its sovereign power in keeping with the express will of the people. In the case of the Fukushima nuclear disaster, wherein the interests of an extended form of sovereign power are not necessarily aligned with those of domestic populations, the government confronts a problem. In order to continue to represent the powerful vested interests of the transnational nuclear village, the Japanese government must either persuade the people to act against their own interests and give their consent to the resumption of nuclear activities, or proceed unilaterally. In short, it must convince or force the people to desire or accept their own continued repression.

Aside from a few notable cases thus far, the Abe government has not reverted to overt physical force to quell popular dissent. This does not mean that its interventions are not violent. Deploying the regulatory tools of a control society, it implemented 'zones of undecideability' under 'special' juridical status during the critical phase of the meltdowns (borders demarcating voluntary and mandatory evacuation which have shifted over time), so that former residents found themselves in limbo with regard to their ownership or the habitability of those spaces formerly known as their homes. Inhabitants were dispossessed and relocated, or not, according to a set of priorities which minimized government liability.

As both TEPCO and the government have been keen to contain expenses and business confidence in investment in TEPCO, nuclear-related corporations, and 'Japan's brand', cost-cutting methods have been widely adopted. The government intervened to protect investors and stakeholders in electricity and nuclear power-related and construction industries by covering the immediate costs stemming from the plant explosions. As it incrementally expanded the mandatory evacuation zones over the days and weeks that followed 11 March, ostensibly to stem public panic, TEPCO and the government identified 'drag-factors' which might have inhibited capital accumulation. In cases removal to 'temporary' accommodation and the return of evacuees, and the sub-contracting of work on contaminated sites to more vulnerable populations including evacuees, the homeless, long-term unemployed, debilitated, the elderly and foreign migrants (Saito and Slodowski, 30 December 2013) people have been exposed to the harsh logic of the economic bottom-line. While keeping electricity production under the control of the Federation of Electricity Suppliers ('*Denjiren-kai*'), major share-holders, financiers and political representatives, and maintaining the corporate structure of the industry, having carried out 'remediation' ('decontamination' and reconstruction) operations for abandoned villages and towns, the government steadily 'incentivised' the evacuees' return.

Thus far, there has been a clear failure to effectively contain or monitor all radioactive contaminants and exposures, whether due to continued leakage of pollutants into the air and ocean, or through the transport, recycling and incineration of radioactive contaminants, through contaminated food and water, and through ineffective decontamination of water and land. The central problem has been that TEPCO, with its corporate ethos, and with government backing, has refused to dedicate more than the bare minimum of its resources to resolving radiation leakage from the ruined plant. Instead, while minimising the damage from legal claims, TEPCO has been permitted to pursue strategies to maximise economic growth. Its neglect for the ruined site, while lobbying to re-start of its other nuclear plants, has been justified in strictly economic terms.

The Liberal Democratic Party (LDP) government, under the Abe administration in particular, has also sought to control the public narrative. Knowledge specialists have sought to 'neutralize' health concerns, by seeking to stamp out 'baseless rumours' (*fūhyō higai*) so as to limit victimisation. The control of public perception through managing information (ideological repetition, marginalising dissentient views, misinformation, surveillance, censorship) has been supported by techno-scientific manipulation (flexible safety standards, inaccurate dosimetry, prescribed medical diagnostics, imposed self-management of health upon workers and civilians through neglect)[i] and has been consolidated, where necessary, by legal bans (*Tokutei Himitsu Hogo Hō* – Special Secrets Protection Law, passed 6 December 2013).

The clearest example of sanctioned misinformation at the highest levels is the 7 September 2013 statement by Prime Minister Abe Shinzō, when he declared to an international audience as part of Japan's bid for the Olympics to the

International Olympic Committee panel in Buenos Aires, that the radiation dispersal from the Fukushima nuclear power plant was 'under control'. Two and a half years after the beginning of the disaster, with the exact locations of the inventory of the Daiichi nuclear reactor still unconfirmed, uninhabitable tracts of land, and continuing dispersal of radiation contamination from the plant and through the circulation of contaminants,[ii] Abe's statement was patently false.

Had Abe's statement been understood as an address to political and business executives to assure them of the profitability of continued investment in Japan may have been closer to the truth. Abe's watchword since being re-elected in 2012 has been to foster national ambition in the 'true spirit of risk-taking and innovation', or 'taking Japan back' (*Nihon wo torimodosu*). Together with the propagated safety myth, Abe's reforms have included 'Abenomics', a casino politics that seeks to increase foreign corporate investment through further deregulation and reduced corporate tax, increased consumption tax, apparent increases in female and foreign workers, large public works projects (such as the 2020 Olympics), and a resurrected and expanded military industry and nuclear power program.

With all of this dynamism, public concern for health effects from the Fukushima Daiichi plant certainly was deflected so as to create an environment conducive to smooth business operations in the private sector.[iii] Another example of overt government intervention in the public narrative was in the coerced self-censorship by Maruyama Hiroshi, the editor of the Shogakukan publisher of *Big Comic Spirits* in which *Oishinbo*, a popular manga series by Kariya Tetsuya, appears. Kariya depicted frank disclosures by public authority figures, who included a town mayor and a doctor, and who spoke of the negative health effects from radiation in Fukushima and from incineration of irradiated materials in Osaka (*Jiji*, 18 May 2014; Osaki, 19 May 2014).[iv] The publishing house stopped the series after receiving complaints from high-level officials including Suga Yoshihide and Abe, as well as from the Fukushima Prefectural government, health officials, town representatives from Futaba, and local individuals from Osaka City and Prefecture.

As nuclear ambassador, Abe's 'under control' statement was driven by the desires of the corporate combines GE/Hitachi, Westinghouse/Toshiba, Areva/Mitsubishi, and directed toward prospective buyers of Japan's nuclear technology in particular. Successful contracts with India, for example, will bring an estimated US $60 billion to the nuclear vendors, excluding cost overruns. As with the deal Japan struck with Turkey in 2014, in India's drive to rival Pakistan, which signed a similar agreement with China to develop fast breeder reactors, Japan may likely be required to tacitly approve the right to nuclear fuel reprocessing and uranium enrichment in the signed agreement. Following similar nuclear agreements concluded with Jordan, Vietnam, South Korea and Russia under the Democratic Party of Japan (DPJ) administrations of Kan and Noda, on 4 April 2014, the Abe government signed off on civilian nuclear accords with Turkey and the United Arab Emirates (JAIF, 19 December 2011; Editor, Japan

Times, 20 April 2014). From the 20 or so countries Abe visited to spruik Japanese nuclear technology, at the time of writing, contracts with six more countries – India, South Africa, Mexico, Brazil, Saudi Arabia, Bangladesh – are under consideration.

While Abe maintains a superficial commitment to the Non-Proliferation Treaty (NPT), his solicitation of a deal with India effectively reversed the 1998 government policy to issue sanctions on India for nuclearizing south Asia with nuclear tests at Pokharan from its clandestine nuclear program. Should Japan criticise India's right to test nuclear weapons or adopt a first-use policy, India would likely cite the 'precedent' established with the US and other NSG members which void such requirements. If Japan objected, India would simply turn to other vendor representatives such as South Korea or Russia.

Abe wooed former Prime Minister Singh with low interest loans for urban infrastructure development in India, a bilateral currency exchange, and a *shinkansen* prototype for use in Delhi. He finessed this good-will with Japan's participation in the US-India Malabar series of joint-military exercises (Chellany, 31 January 2014). In return, the former Prime Minister Singh publicly declared that 'the safety and livelihoods of people would not be jeopardized in our pursuit of nuclear power' (Editor, Dianuke, 5 February 2014). The in-coming Hindu nationalist government of Prime Minister Narendra Modi, with his declared intention to distribute electricity to all communities in India, embraced both solar and nuclear energy. Modi took up Abe's initiative in 2014 by appearing to remain committed to India's Civil Liability for Nuclear Damage Act (2010) in which liability is taken on by vendors and component manufacturers (Bhattacharjee, Sasi, 12 June 2014), while also organising a consortium of insurers in the event of mishap including the Nuclear Power Corporation of India (NPCIL). At the time of writing, Modi plans to finalise the Japan-India Nuclear Cooperation Agreement in Japan on the 1st of September 2014.

In addition to Japan, Australia has also chosen to contribute to India's nuclear ambitions. Following the overturning of the ban on selling uranium to India under the Gillard government in 2012, Prime Minister Tony Abbott will sign an agreement with India in September 2014. Where Japan will supply the turbine required for 1000 MW capacity and further exploit potential in India's nuclear sector, Australia, along with other suppliers, will also be able to exploit India's increased need for uranium as it builds roughly 20 new nuclear reactors over the coming years.

These two Agreements, which are by no means simply coincidental, will enable India to use nuclear reactors and fuel for its civilian nuclear energy program, while claiming, as a non-signatory to the NPT, the right to indigenous nuclear weapons production

In this sort of nuclear power-bloc building, sovereign power concretises the abstract and abstracts the real. Along with the Olympic Games, the spirit of patriotic cooperation in adversity, 'creative' wealth-recovery programs have been deployed that include optimistic tourism and local produce drives for Fukushima Prefecture, "decommissioning and remediation", or energy recovery schemes in affected areas (renewable, waste reprocessing, mining).

If this situation can be framed as 'war', as some are want to do, then it is a covert or silent war in which mobilised labour and citizens who continue to be locally exposed to radioactive contaminants, will bear the greatest burden as collateral damage in the real violence of capital accumulation by the transnational nuclear village within extended sovereign power (see Klein 2007, 256-8; Harvey 2003, 127-73).

Re-Constituting the Sovereign

The response of sovereign power to the eruption of the nuclear disaster at Fukushima Daiichi has also served to highlight the changes underway in the geopolitical landscape of the East Asian region which can be understood in a legacy stretching back to 1945.

When Abe Shinzō was returned as Prime Minister for the LDP on 20 December 2012, he declared that 'Japan was back' and that it would never be 'a two-tier nation'. Claiming to be a patriot who was 'doing what is right in order to protect people's lives' (Gibbs, Beech, 2014), with majority control, the Abe faction sought to further escalate Japan's militarization. It further hollowed out the constraints imposed by the 1947 Constitution in which Article 9 renounces the use of force to settle international disputes and defines Japan's passive defence strategy (*senshū bōei* or *judōteki na bōei senryaku*).[v] Article 96 requires that any Constitutional amendments are to be initiated by the Diet and approved by a two-thirds majority in each House, and then ratified by the people through a referendum or special election conducted for that express purpose. Instead, the Abe faction sought discretionary power to interpret the Constitution and where necessary to declare a 'state of exception,' in which a legal suspension of the rights of citizens permits extra-judicial sovereign intervention in the national interest.

'Taking Japan back', was not just to return to but also to extend the policies of Abe's grandfather Kishi Nobusuke. Following his release from Sugamo prison as an unindicted Class A war criminal, Kishi was critical of the Tokyo War Crimes Tribunal (1946–48). After becoming Prime Minister in 1957, in his negotiations with the then Ambassador Douglas MacArthur III to formulate the US-Japan Mutual Security Treaty in 1960, Kishi sought closer but more symmetrical military ties with the US. As the former Commerce and Industry Minister of Imperial Japan's wartime cabinet under Tōjō Hideki, and as a key CIA operative (Weiner 2007, 116-21), Kishi used his experience and connections to mobilise the transformation of Japan into a construction nation (*doken kokka*) and propel it toward a plutonium economy and eventual global economic superpower status in the 1980s.

Whether romanticizing or purifying the criminality of Imperial Japan's past military aggression, the Abe faction actively sought the return of the emperor's status from 'symbol' (*shōchō*) to 'head of state' (*genshū*),[vi] and regenerated the militarist tradition of honouring the sacrifice of all troops in the name of the Japanese Emperor since the Meiji Restoration. The highly public and semi-official visits to the Yasukuni Shrine (roughly 150 lawmakers have visited since Abe's initial visit as Prime Minister on 26 December 2012), have been defended as an exercise of the 'freedom of belief' (Lies, 22 April 2014).[vii] This is part of a practice of 'beatifying' the historical narrative which has been ongoing since the Nakasone administration, wherein the national education curriculum is purged and official apologies are moderated with regard for the systematic massacres and use of forced labour carried out by the Japanese military during the Fifteen Years war (including the system of 'comfort women' stations of forced prostitution). This also includes making the national flag and anthem compulsory in Japanese schools. The refusal to extend apologies or reparations to individual victims (and their families) is based on the claim that these issues were resolved in the 1951 San Francisco Treaty, the bilateral 1965 Japan-South Korea Normalization Treaty, and the 1972 Joint Communiqué of the Government of Japan and the People's Republic of China.

To manufacture consent for this systematic program to 'escape the post-war regime' (*sengo no dakkyaku*) and 'normalise' the nation from its 'masochistic' post-war historical narrative, the Abe faction has re-appointed the governing body of the national broadcaster NHK. Without this major shift in public consciousness so as to recover lost patriotic pride, younger Japanese presumably could not assume the 'normal' predisposition to self-sacrifice for the nation, and the Abe faction could not cement these reforms in a 'reinterpreted' Constitution. The Abe faction, ironically at the urging of US officials, also established the National Security Council (NSC) in early December 2013. Mirroring the US model, a Bureau of sixty officials coordinating defence and foreign ministries, and a dozen Self-Defense Force personnel, the NSC operates to reinforce executive control to enact the right to collective, and possibly unilateral, 'self-defence' in 'emergency situations', the right to manufacture and sell weapons overseas, and military and police powers of surveillance, and the legal obligation of citizens where required to defend national sovereignty and independence through the preservation of national territory, air-space, and access to its intelligence and resources. Reminiscent of the Peace Preservation Law (1925) which extended discretionary power to the *Kenpeitai* secret police, the vague and broad Secrets Law rammed through Parliament in the middle of the night denies public access to certain information in perpetuity, and stipulates punishments from five to ten years imprisonment for offenders (who disclose such information) (Torres, 5 December 2013; Repeta, 10 March, 2014).[viii] Negotiations for transnational information sharing between the NSC and the United States, Britain, Australia, New Zealand, and Canada, and certain EU members are underway.

While assuring the peaceful purposes of these fundamental changes, the Abe faction has sought to realize a dream (held by Kishi, Yoshida Shigeru, Nakasone Yasuhiro, and others) to return Japan to 'great power' status previously legitimated under the Meiji constitution, with which it can shape the international security structure. Given that a middle power is reliant upon its economic weight to exert influence rather than possessing the weaponry required to enforce its threats and agreements (Suzuki, 2006, p. 192; Mearsheimer, 2001, 382, 399), these changes are contributing to the destabilization of official relations with South Korea, China, Taiwan and Russia in East Asia.

Abe's assertion in early 2014 that 'Japan is back and thriving, and that its return is indispensable for global stability and prosperity', was indicative of the intention to enable Japan's capacity to engage in foreign wars. This was confirmed on 1 July 2014, when the 'right to interpret the Constitution for collective security measures' was passed in the Diet.

Strategic Re-assessments in the Security Environment

The suppression and neglect of the consequences from the Fukushima nuclear disaster can be viewed in the context of this renewed military and economic push by the Abe government. The most commonly cited reason for the radical political shift in Japan's orientation adopted by the Abe faction is China's rapid high economic growth including its ownership of large number of US government bonds, its 'double-digit' military expenditure and capacities, and its accumulation of global resources.

Since the first Persian Gulf War of 1991, Japan's Self-Defense Forces (JSDF) have shared integrated weapons technology and participated in 'regional deterrence' exercises in the region. In its joint-operations with the US in the second Gulf War, JSDF ground troops have been deployed to Iraq, and its navy (MSDF) has assisted in re-fuelling for the war in Afghanistan. The JSDF has also been deployed in disaster operations in Japan, Haiti, Indonesia, and the Philippines. For all his desire to shed the post-war system, in the context of conflicting claims over the rocks, islets, reefs, shoals and fishing grounds in the East and South China Seas (which include the Senkaku/Diaoyus, Dokdo/Takeshima islands, the Paracel islands and Spratly islands), Abe assiduously emphasised Japan's position in the post-war order under the US hegemon.

This American-crafted order was established under the San Francisco Treaty system, reflects a raft of US foreign policies and bilateral military and trade alliances that suit American interests, including the right to station military bases on foreign soil and the right of passage for its blue-water navy. Historically, those nations that have not supported the demands of these hegemonic conditions have not been privy to preferential trade or military agreements.

The US 7th Fleet, for example, has exercised control over large areas of the Pacific Ocean, including bilateral military alliances and chains of military bases and Pacific ports, shipping lanes and maritime choke-points, with which to ensure safe passage for the goods of its favoured multinational corporations and to patrol the borders of its favoured partners (Reed, 20 August 2013). The policy trajectory taken by Japan's leadership over seven decades since the end of WWII has been directly affected by this security regime. The Ryūkyū and Ogasawara (Bonin) islands, for example, were sacrificed to US military rule in exchange for a privileged military and economic relationship with the hegemon.

The US has been a staunch defender of Japan since 1945 per its security treaty to maintain advanced forward deployment on the archipelago. The US also long advocated the 'trilateral defense cooperation' with Japan and South Korea, to lessen its military burden in the region while continuing to maintain the capacity to contain China. In the 1970s, however, the US also supported China in their shared interest to contain the Japanese economic 'wild horse', as Zhou En Lai put it (Schaller in Leffler and Westad, 2014, 174). In this sense, US policy in East Asia has been one of double containment. Unresolved tensions in Japan–China relations are to the advantage of the US, as they position it as arbiter in conflicts between China and Japan. This entrenches the deep division between them. And while both Japan and the US have been economically dependent on China's rise over recent decades, Japan's relationship with China has been carefully circumscribed so as not to undermine American dominance in the region.

It is fairly clear from the historical legacy of bilateral agreements with the US in the Asia-Pacific, including Japan, South Korea, Taiwan, Australia, the Philippines, Singapore and Vietnam among others, that the intention is to form military-economic alliance structures which control the capacity of China, and Russia in the central Asian domain, to form a Eurasian power bloc.

In seeking to protect its economic zones on par with other nations, like Japan and South Korea, China established its own Air Defence Identification Zone (ADIZ) in 2013. It also claimed areas of the Spratly and Paracel Island clusters for mining and sealane protection, to which Vietnam and the Philippines responded by striking 'strategic partnership' agreements with the US and Japan.[ix] Japan also forged stronger ties with the Association of Southeast Asian Nations (ASEAN) in mid December 2013.

Over the course of a year, public fear has been fomented of the spectre of an aggressive Chinese behemoth and an unpredictable and provocative North Korea (and its imminent fourth nuclear test). Following an illegal *coup d'état* by NATO and US-backed militia in Ukraine in February 2014, to precipitate the inclusion of Ukraine in the EU, Abe and Obama held a summit in May 2014. They declared that any challenge to upset the US-Japan centred post-war order would not be tolerated. Then Abe visited six nations in the European Union. The consistent message from each of these meetings was to condemn violations of international law by intervention in the territorial sovereignty of others. The consensus was to impose economic sanctions on Russia for its intervention in 'Ukraine's territorial sovereignty' by facilitating the secession of Crimea to it. But the Abe faction also used the opportunity to warn China and, less directly North Korea (Reuters, 24 April 2014).[x]

Abe and EU leaders agreed to share 'defence equipment', weapons research and manufacturing, and information security technologies, which include unmanned (drone) surveillance submarines (with France), renewable energies technology (with Germany), nuclear reactors including fast breeder technology (with Britain and France), and military logistics for 'peace-keeping and humanitarian military missions' (JSDF and British military) (Aoki, 13 April 2014).

In response to these manoeuvres and in light of the Ukraine crisis, China and Russia, normally rivals, initiated greater cooperation through an independent joint ratings agency. With an apparent new Cold War stand-off in central Europe and the possibility of conflict in East Asia, it has been easier for Abe to claim that Japan must prepare for 'collective security' and 'peace-keeping' overseas.

In the US tradition of justifying large-scale military intervention through dubious pretexts, including 'pre-emptive regime change' – the Tonkin Gulf incident in Vietnam, the WMD in Iraq, the use of sarin by the Syrian Government on its own people (Hersh, 17 April 2014, 21-4), the staged insurrection in Ukraine[xi] – it is not unforeseeable an incident could precipitate a conflict in which Japan would be called upon. Should an over-extended and financially challenged US protector be diverted, however, the realists argue that Japan may need to conduct an independent power struggle with China (Campbell and Sunohara, in Campbell et al., 2004).

Japan's recent lift on the ban on the right to sell weapons extended its pre-existing export of 'dual use' components for military use, and its highly advanced military capacity. With contracts for new aircraft, naval and urban deployed ballistic guided-missile technology, and joint-operated bases at Henoko on Okinawa and Yonaguni in the Ryūkyū chain, Japan is primed to become a major competitor in the global weapons industry. This military augmentation to the economy will help accrue the greater geopolitical influence the Abe faction seeks for Japan. At the same time, such high stakes drama marginalises the apparently dull crisis at Fukushima Daiichi.

From Energy Security to Nuclear Security

'Self-restraint' has long been the preferred term for Japanese official policy on the possession of nuclear weapons (Soeya in Alagappa, 1998, 228–33; Kamiya, 2002/03, 65–7). With the Japanese public already concerned about the security environment and international order, removing the brakes on nuclear-armament may even be perceived as bringing greater stability to the region. This often over-looked factor is another reason for the stubborn commitment of the Abe government to Japan's nuclear program and neglect for the ongoing nuclear disaster at Fukushima Daiichi.

A strong and popular anti-nuclear weapons movement in Japan arose from the nation's experience of the atomic bombings of Hiroshima and Nagasaki and the *Fukuryūmaru No. 5* incident in 1954, in which Japanese fishermen were exposed to fallout from a secret hydrogen bomb test at Bikini Atoll. As discussed below, Japan's 'non-nuclear policy' (*hikaku seisaku*) was adopted as part of its negotiated military and political position as a 'peace state' in the post-war system. As the generation which engaged in this anti-nuclear campaign ages, realist security calculations have begun to succeed in eroding this status as a non-nuclear weapons state.

In the push for nuclear power by the US Atomic Energy Commission after 1945, under the influence of Henry Kissinger whom he met at Harvard, a young Nakasone Yasuhiro urged the Diet to introduce a budget for nuclear energy production and research. Aligned to the US nuclear production model and its reactor designs, and the tenets of the Baruch plan (1946) and the Atoms for Peace campaign (1953–55), Japan was one of the first nations to embrace commercial nuclear energy production. It passed Japan's Atomic Energy Basic Law (1955) and stipulated the development of nuclear energy production for peaceful purposes only. Japan joined the International Atomic Energy Agency (IAEA) when it was formed in 1957, the same year it started its first experimental nuclear reactor. In keeping with the US ploy to control the international distribution of nuclear related materials within its geopolitical orbit, the IAEA (1957) was mandated to promote nuclear energy production and to ensure, albeit with no enforcement powers, that nuclear reactors were for the research, development, and utilisation of the peaceful production of atomic energy.

Like many countries, Japan tailored the IAEA regulations to its own interests. China also ran a uranium enrichment plant with two hydroelectric power stations built on the Yellow River in 1958 to produce an enriched uranium bomb for a successful nuclear test on 16 October 1964. In December 1964, Prime Minister Satō Eisaku, brother of Kishi, expressed to US ambassador Reischauer that it was common sense that Japan possess nuclear weapons (Gavin, 2004/5, 117, fn. 50). This provided the incentive for President Lyndon Johnson to reaffirm in a joint communiqué with Satō the United States' commitment to the Security Treaty to 'defend Japan against any armed attack from the outside' (Akiyama, in Self and Thompson, 2003, 81–5; Japan Defense Agency, 2006, 95, 142). At that time, the US did not want Japan to be nuclear armed, if it could manage it.

In 1967, a year after the first commercial nuclear power reactor went on-line near Tōkai Village, Ibaraki Prefecture, Satō announced the 'three non-nuclear principles' – not possessing, not developing, and not introducing nuclear weapons into Japan. Despite a flurry of internal debate over nuclearization, Satō officially committed Japan to reliance upon US nuclear deterrence in 1968. At the same time, while it promoted the peaceful use of nuclear energy, in September 1969, the Satō administration drew up guidelines to maintain the economic and technical potential to convert its nuclear energy technology into nuclear weapons (Yamada, 25 March 2012). Even as it signed the Treaty on the Non-Proliferation of Nuclear Weapons (NPT) in 1970 (Endicott, March 1977, 275–92), the same year the Fukushima Daiichi plant was completed, it was far from clear whether it still intended to develop tactical nuclear weapons. Although the non-nuclear principles were adopted by the Diet in 1971, and have stood as testament to Japan's restraint through successive administrations, these principles were never legislated. They remain contingent upon prevailing conditions and can be changed at will. When Japan finally ratified the NPT in 1976, its nuclear policy was framed in four pillars: the three non-nuclear principles; nuclear disarmament; reliance on US nuclear deterrence; and the peaceful use of nuclear energy.

Calls from the US for Japan's greater participation in regional security management can be traced back to the 'Reverse Course' doctrine of 1948 and John Foster Dulles' re-set of the US-Japan relationship to one of sovereign equals (Dulles, 1951). The then-Vice President Nixon's in 1953 also stated that it had been a mistake to advise 'Japan to forego the maintenance of a military force' (Dionisopoulos, 1956, 445, fn. 33). Fielding a military force and nuclearization were regarded as different issues, however, and Japan was discouraged from obtaining a nuclear capacity (as was South Korea).

Not just to avoid sparking an arms race in East Asia, this was to maintain the 'balance of power' over which the US presided. For the most part, Japan's 'special' role as a carrier and facilitator of US military bases and actions in the region hinged upon credible deterrence from a US nuclear umbrella to 'balance' nuclear-capable China, the Soviet Union, and later, North Korea. Under this umbrella, the Japanese government steadily augmented its conventional defence capabilities, while it preserved the 'peace state' illusion for assuaging the public concern for the nation's re-militarisation.

The fact is that Japan has long considered and made preparations for the day when it might possess indigenous nuclear weapons. Prime Minister Kishi's statement to the Diet in May 1957 that it was not unconstitutional to acquire tactical nuclear weapons as a 'minimal' requirement for the defence of the nation, has been used by defence strategists who claim that Japan's non-nuclear status is not a viable national security option. Kishi argued that such a situation was possible with nuclear-armed China or Russia, and also in the event of Communist 'human sea' tactics (invasion by large conventional forces) (Office of Intelligence Research, 2 August 1957, 2).

The spectre of Japan's nuclearization has often been raised as a way to stimulate public debate on national defence. Even if nuclearization was ruled out, these public debates often reached a compromise for the development of conventional weapons within the Japan-US military alliance structure. In short, the LDP-dominated political scene since 1955 has seen steady re-armament over three phases beneath a veil of popular pacifism. Even when the military budget was limited to 1 percent of GNP following the oil shock of 1973 as announced by Prime Minister Miki Takeo in 1976, Japan's rate of military expenditure was the highest in the world by 1970. Japan's overall economy had grown so vast that it neared the top ten nations in military expenditure by the late 1970s (Welfield 2013, 364–70). The Nakasone administration then exceeded the 1 percent limit in 1987.

After ratifying the NPT in 1976, the oil shock was used as a pretext in resource-poor Japan for nuclear energy development. Japan also resumed its fourth long-term military build-up amid the arms race of the second Cold War. Policy makers devised a 'dual-use' potential by planning to 'close the nuclear fuel cycle'. This meant nuclear fuel reprocessing and fabrication of plutonium produced from fast-breeder reactors and the use of MOX (uranium/plutonium mixed oxide). In producing more plutonium than was consumed, the idealized notion was of an endless and self-sufficient fuel supply. This also offered the potential to divert plutonium toward weapons-grade material.

In February 1977, the same year the Jōyō Fast Breeder Reactor (FBR) in Ibaraki went critical, US Vice President Walter Mondale met with Prime Minister Fukuda Takeo to discuss his concern for Japan's intentions. The Ford and Carter administrations had rejected the idea of spent fuel reprocessing for energy production due to plutonium stockpiling and likely diversion and proliferation into nuclear weapons production. In 1978 and 1982, the Cabinet Legislative Bureau re-confirmed that nuclear weapons possession would be constitutional if it could be proven that they were the 'minimal level of armed force necessary' for 'self-defence'.

In the 1980s, realists in Japan argued for a Gaullist posture to achieve parity with the Soviet deployment of SS-20 intermediate-range ballistic missiles (IRBMs). When Cold War tensions subsided and Washington and Moscow agreed to eliminate IRBMs, they became concerned that the dissipation of antagonism toward a Communist bloc would weaken the justification to the American population of US blood and treasure for the continued defence of Japan. With the loss of a credible US nuclear deterrent due to a potential US military draw-down in South Korea following a 'non-aggression' pact with North Korea, they argued that Japan's nuclearization would be essential (Nakanishi, October 2003, 52). In the late 1980s, the Japan Nuclear Fuel Services (JNFS) began construction of a nuclear reprocessing plant and storage facility on land bought near Rokkasho village, Aomori Prefecture for the production of enriched and depleted uranium.

In 1992, following international criticism for Japan's lack of participation in the Persian Gulf War, a *Yomiuri Shinbun* study found that the failure of the NPT in verifying weapons of mass destruction in Iraq was a failure of the NPT. As its concerns were more for North Korea and China than Iraq, it recommended the inclusion of a 'no-WMD clause' in the non-nuclear principles. In July 1992, instead of sending spent nuclear fuel to Sellafield and Cap le Hague, the plan was formally announced to reprocess spent nuclear fuel from the cores of Japan's FBRs (Jōyō and Monju) at the Tōkai Recycle Equipment Test Facility (RETF) in Ibaraki operated by the Power Reactor and Nuclear Fuel Development

Corporation (PNC). The purpose was to reprocess a small quantity of high-grade plutonium (98 percent Plutonium 239).

In December 1992, official permission for the construction of the 'commercial' (commercially non-viable) Rokkasho Reprocessing plant was also granted. Editorials in the *Nikkei* and *Korean Daily* newspapers at the time recognised this as part of the completion of Japan's 'transition to a political and military superpower' (Tsuchida, 1992, 3, 6). South Korea responded by procuring two Canadian CANDU reactors, which can produce weapons-grade plutonium. In September 1992, China conducted its first test of a tactical nuclear weapon. In the same year, Russia committed to the construction of four FBRs, and the US continued to develop smaller nuclear warheads. In May 1993, North Korea test fired a Sodon 1 missile into the Japan Sea. By this stage, it was clear that a nuclear arms race had already begun in North East Asia.

In 1994, citing Japan's plutonium production program, North Korea withdrew from the NPT. This year also marked the beginning of Japan's active contributions to UN initiatives and resolutions for the elimination of nuclear weapons. It followed up with strong protests and sanctions for nuclear tests by France and China in the mid-1990s. In 1995, in a bid for greater influence, the Japan Defence Agency (JDA) formally called for Japan's permanent status at the United Nations, despite being a non-nuclear weapons nation. In 1996, it signed the Comprehensive Nuclear Test Ban Treaty (CTBT), despite resistance from the US. At the same time, it also signed the 'US-Japan Joint Declaration on Security' which provided for an expanded role for the Japanese Self-Defence Forces beyond the 'Far East' in cooperation with the US. Japan also froze aid to India and Pakistan after their nuclear tests in 1998.

After 1998, realists in Japan claimed that the complete eradication of nuclear weapons technology was impossible, and prescribed capabilities to match China to fill the power vacuum left from a projected US withdrawal from Asia (retreating to Guam and Hawaii). They indicated the need to develop a blue-water navy capacity to meet China's ability to patrol the borders around Okinawa and the Senkaku/Diaoyu Islands in the East China Sea.[xii]

The North Korean ballistic missile tests over Japanese air space in 1998 and terrorist attacks on the United States on 11 September 2001, combined with the failure of Prime Minister Koizumi's negotiations with North Korea over the abduction issue and discontinuation of its nuclear program, sealed the end of the 1994 United States and North Korea Agreed Framework. Realists took the opportunity to forcefully argue for more proactive collaboration with the US. It was claimed, by the LDP cabinet secretary Fukuda Yasuo in 2002 and Prime Minister Abe Shinzō in 2006, that tactical nuclear weapons possession was constitutionally legitimate, that it was necessary for minimal self-defence and that it was not offensive to other nations. (Furukawa in Self and Thompson, 105; Kyodo, 14 November 2006).

It is often stated that Japan's nuclearization would upset the geopolitical balance through the abandonment of nuclear arms control in the region and would erode the NPT and IAEA regimes. But Japan has nearly completed the process without sanction. While a credible intent to nuclearize has apparently retained Washington's commitment to defend Japan against China and North Korea, and permitted it to bolster its conventional capability, the fact is that the US may not necessarily disapprove of Japan's nuclearization.

In June 2002, the US had already announced its withdrawal from the Anti-Ballistic Missile Treaty (ABM) of 1972. In 2003, to counter a perceived risk of self-isolation, the US Vice President Richard Cheney and the journalist Charles Krauthammer both suggested that Japan's nuclear weapons could be supported if its nuclear policy complemented US deterrence (Krauthammer 3 January 2003; Mochizuki, July 2007, 314; Frum, 2006, A25). Whether sharing tactical nuclear missiles with the US, as exists in Europe, or fueled with indigenous plutonium, with US permission it was increasingly possible for Japan to obtain tactical nuclear weapons.

On 9 December 2003, in support of the US-led wars in both Afghanistan and Iraq, Prime Minister Koizumi announced that Japan would dispatch the JSDF. In 2004, in continued focus on North Korea, the *Yomiuri Shinbun* called for both a no-WMD clause and a national military ('*guntai*' as opposed to '*jieitai*' or 'defensive' forces). In the same year, US President George W. Bush deployed the Aegis Anti-Ballistic Missile Defense (ABMD) system after a decade of development initiated by President Reagan's Strategic Defense Initiative (SDI). Ogawa Shin'ichi of the National Institute for Defense Studies (NIDS) commissioned by the former Defense Minister Ishiba Shigeru, claimed that 'counter-offensive' conventional interception of missiles heading toward US naval vessels near Japan or toward US territory would only be possible if the Constitution was altered (Mochizuki 2007, 318).

In early 2006, Bush obligingly declared the 'United States has the will and the capability to meet the full range of its deterrent and security commitments to Japan'. Japan joined the US Missile Defence (MD) program in that year and contributed to its technical development (R&D) (Furukawa 2003, 120). A US Aegis destroyer equipped with the ABMD system was deployed to alleviate 'security abandonment' concerns near Japan. The Japan Defense Agency subsequently announced a joint bilateral station for interceptor missiles in Nagasaki (Twomey in Self and Thompson, 2002).

North Korea answered by resuming missile tests in July 2006. On 5 September 2006, an Institute for International Policy Studies report commissioned by Nakasone Yasuhiro proposed that Japan's defensive status would be altered to permit offensive force in lieu of 'an international security shift' caused by an imminent security threat (such as North Korean nuclear tipped missiles). In this logic, 'pre-emptive' strikes on enemy bases was an act of self-defence and necessitated rationalizing US bases and weapons in Japanese territories in the case of a lack of plausible US nuclear deterrence (IIPS 2006, 6; Japan Times, 6 September 2006). This was followed by the first nuclear test by North Korea on 9 October 2006.

Japan imposed sanctions on North Korea (Nakata 14 October 2006; Yoshida and Ito, 2 November 2006), North Korean imports, vessels and citizens from entering Japanese ports. Japanese realist strategists and policy makers repeated that a nuclear option should not be ruled out should North Korea advance from one test to full-scale nuclear weaponization and China continue its military build-up. In late November 2006, the Defense Minister Kyūma Fumio stated that Japan should relax the third non-nuclear condition so as to allow for US ships carrying nuclear weapons to pass through Japanese waters and bolster US extended deterrence (Yamaguchi 26 November 2006; Kyodo, 21 December 2006; Editor, Yomiuri, 21 March 2007). Based on a precedent set by a secret agreement in 1968 between Satō Eisaku and Vice President Nixon to permit the US to station nuclear weapons on Okinawa and at the Yokosuka naval base, this view carried consensus.[xiii] At best, it is true that Japan only maintained '2½ non-nuclear principles'. The Japan Defense Agency was upgraded to the Ministry of Defense in January 2007.

Present Realist Assessments

According to the NPT, if Japan were to obtain nuclear weapons it would be required to return all nuclear materials 'to the original exporting country'. Sanctions by its uranium suppliers could be expected (US, UK, France, Canada, Australia) (Hisane 2006). The NPT is not sacrosanct, however. As seen in the US–India energy agreement of July 2005 and US–India Civil Nuclear Agreement of October 2008, the decision of the NSG (including Canada and Australia) to export uranium to India as a non-NPT signatory, and US and NSG policies which favour Israel, a non-NPT state in the possession of 200 to 400 nuclear weapons, and do not favour Iran, which has none, there is no equivalence in application (Tellis 2005, 33; Potter 2005; 343-54; Blix, 12 May 2014).

Following the Nuclear Security Summit in The Hague in March 2014, after initial resistance, Japan agreed to return 320kg of weapons-grade plutonium (roughly 110 bombs) and 200kg of highly enriched uranium to the US. This material was apparently originally obtained with the purposes of developing a closed fuel cycle.

The expressed concern was for potential terrorist interception of these materials. Yet, this overlooked Japan's already existent stockpile of roughly 44 tonnes of plutonium, the fourth largest stockpile of 'civilian' plutonium. Somehow it was also missed that the Rokkasho Reprocessing facility currently stores most of the reprocessed spent fuel from the nation's reactors (currently reprocessed in the UK and France) (Okuyama, 13 April 2014; Johnston, 15 November 2005; Johnston, 6 January 2006), and is intended to produce 800 tonnes of uranium and 8 tonnes of separated plutonium annually. Iran, an NPT state since 1970, is still negotiating for the right to produce less than 1kg of plutonium from its reactors (Haaretz, 12 June 2014).

Despite years of mis-management of these nuclear plants, the Abe administration's comprehensive energy plan includes the use of stored plutonium in nuclear recycling and pluthermal programs (use of MOX fuel in power reactors). This will increase Japan's already significant plutonium inventory. Unlike France, Britain and Russia, all of which reprocess plutonium for nuclear fuel, Japan is the only NPT non-nuclear weapons state to have been granted permission in the Japan-US nuclear cooperation agreement to reprocess 'US origin' fuel, which is its total inventory as it includes any fuel irradiated in US-made reactors (Pomper and Toki 2013). With both Rokkasho and the Monju FBR operational, and with fuel fabrication by no means beyond Japanese technicians, the weapons potential from the current stockpile is conservatively estimated at 1000 warheads, and could amount to as many as 5000 warheads (based on the IAEA estimate of 8kg of plutonium per nuclear weapon) (Johnston, 10 May 2014).

China's sustained contention that Japan's stockpile of nuclear materials was of 'grave concern for the international community' appeared again when it complained that Japan had not reported 640 kilograms of plutonium to the IAEA in 2012-3. The Japan Atomic Energy Commission stated that this was deemed exempt from reporting requirements because it was for the mixed plutonium-uranium oxide fuel program at the Genkai nuclear plant in Saga Prefecture (Ramzy, 10 June 2014).

Japan's consistent refusal to abandon its full-scale fuel fabrication and reprocessing program of spent fuel from light-water reactors (LWRs) indicates its dedication to closing the nuclear fuel cycle. China's concerns are that establishing an indigenous energy supply in this way would make Japan invulnerable to NPT sanctions. It could then divert plutonium toward deployment on nuclear armaments. This would overcome its reliance on US nuclear deterrence, should this be deemed necessary, and allow Japan to assume greater geopolitical authority in a changing order (Tabuchi, 9 April 2014).

Strategic realists in the US and Japan have both exacerbated and taken advantage of hostile relations with North Korea and China to justify expanding their joint security regime and its associated industries. While the US is apparently concerned to avoid nuclear proliferation, Japan is apparently concerned that US military draw-down could make it more vulnerable (Smith, 7 October 2013; Nishihara, 14 August 2003; Teshima, December 2006, 116–7; Nakanishi, December 2006, pp. 60–1).

The shift of US military posture toward missile and satellite weapons over the last two decades, has seen the restart of the US Missile Defense (MD) program in 2003, and the participation of the Japan Self-Defense Force in research into Anti-Ballistic Missile Systems (ABMS). These use a multi-layered 'fence' of SMIII and PAC-3 missiles for missile interception in coordination with US nuclear and conventional extended deterrence (Dawson, 9 December 2012). This system is to 'neutralize' or disable enemy nuclear weapons at specific points in their flight or before they are launched.

With the newly founded NSC, the intention to 'upgrade' to pro-active status with the US (on par with Britain) was unveiled. As part of collective defence, the ABM system is networked with F-35 planes, a new fleet of aircraft carriers and possibly nuclear submarines, and satellites in real conditions (Defense Industry Daily, August 2013; Bender, 6 May 2014; Okazaki, 23 April 2006; Jimbō in Self and Thompson 2003, 33, 41–4). In the Constitutional adjustment to an unequivocal 'self-defense military', this system is designed for the (offensive) force beyond defence of the homeland which will fill the supposed vacuum left by the deployment of the US military elsewhere.

Should Japan decide to nuclearize these missiles, nuclear reprocessing is critical to produce high-enough grades of plutonium or enriched uranium for deployment on small warheads which can be delivered by mobile platforms that can escape detection. Japan's restraint on doing so is contingent upon there being no radical change that threatens Japan's 'national interests'. A reunified Korea armed with nuclear weapons that was hostile to Japan and/or aligned with China, for example, or a strengthened military and economic alliance between Russia, China, North Korea, and possibly Iran, would be enough for the realists to finally achieve their dream. It may take far less.

Given the aggressive interventionism of US foreign policy to date, and Japan's role as a pivotal base in US Pacific Command strategy, it is plausible an incident could be used to justify pre-empting or destabilizing Eurasian alliances in the near future. With enhanced executive power cemented in the constitution, the LDP could obtain the requisite votes in the Diet to possess nuclear weapons. Japan would then seek a permanent position on the UN Security Council for greater international leverage, and, with the backing of the US, it would engage in earnest in a nuclear arms race with China, South Korea and Taiwan, and possibly the Philippines and Vietnam.

Conclusion

While it was obvious that Abe's claim that the conditions at the Fukushima Daiichin Nuclear Power Plant were under control was disingenuous, it has been less clear that material conditions have been abstracted for the protection of transnational investors' confidence in these sovereign nuclear assets. It is increasingly obvious, however, that the equation which calibrates corporate rights to ownership and profit-making as non-negotiable while continuing to sacrifice human rights to a clean environment, health and suitable living conditions must be inverted.

A false division between nuclear weapons and civilian nuclear reactors has been institutionalised since the Atoms for Peace campaign. This promoted a dual-use program of nuclear technology – atomic weapons and nuclear energy production. Under this system, nuclear weapons development continued while civil nuclear technology was promoted as a techno-utopian, commercial elixir to the world's problems. If Japan's nuclear power program was singularly commercial, then the huge costs and complexities in its centralized structure and entire fuel cycle (from mining to waste storage, and the construction, maintenance and decommissioning of plants), makes the nuclear enterprise far less profitable than any other form of electricity generation.

The disaster at Fukushima Daiichi has demonstrated that should another nuclear disaster in Japan or in a client country occur, neither the Japanese government nor its corporate vendors could or would compensate for the actual toll from the dispersed radioactive contaminants. Instead, they would seek to rely on contractual waivers in which the host government would have to intervene (as in the US Price-Anderson Nuclear Industries Indemnity Act, 1957, or the new NFIL agreement in India).

The invidious distinction between commercial and military nuclear power occludes the symbiosis of Japanese and American nuclear corporations and their respective military industries. It also conceals the fact that other countries conscientiously import nuclear technology, as in the case of India, the UAE or Turkey of Japan's nuclear technology, for possible nuclear weapons production. The double-standard is that other nations (such as Iran or North Korea) are sanctioned should they adopt similar practices.

In light of the demonstrated will to spend on new military hardware, weapons innovation, nuclear programs and public relations programs, the miserly neglect of radiation contaminated areas is even more obnoxious. Despite its advanced economic status, the Abe government seems incapable of taking appropriate and responsible measures to protect people in affected areas (not a small area) – closing all schools; providing the costs for evacuation re-housing and re-habitation; halting domestic and international consumption of contaminated Japanese products; providing national medical treatment commensurate with the actual toxicity of radiation exposure; storing all radiation contaminated materials until satisfactory disposal methods are found; assisting in a coordinated international effort to 'remediate' the ecology of the Pacific Ocean affected by ongoing contamination by radioactive waste.

As the Abe faction exacerbates threats so as to validate its desires for great power status, the corollary is ignored that North Korea, often absurdly aggressive, has been pushed to develop nuclear weapons as a bargaining tool. Even in limited and 'pragmatic' threat calculations – whether China matches or surpasses either Japan or US-Japan joint capabilities in and around the Japanese archipelago – Japan's military and technological capacity emerges far superior (Gurtov, 10 March 2014). China's nuclear missiles cannot reach American centers. China's nuclear counterforce capability does not match the US equivalent. China does not possess a credible second-strike capability (it maintains a no-first use policy and no use against nuclear-weapons-free nations and zones). China's reliance on 'dual-use' technology from France, Germany, Britain, Russia and the Nederlands for military-ready weapons makes it vulnerable to embargo and systems obsolescence (Lague, 20 December 2013). Despite its much touted economic rise, China's military budget is dwarfed by the US, which has the biggest military budget by far.

Whether in a nuclear exchange, or from deliberate attack, natural disaster and/or negligent operation of a nuclear plant, security assessments of the residual effects of radiation contamination on human populations are inadequate. (i.e. China has more sparsely populated land and therefore has the advantage over Japan with its densely populated Japanese archipelago) (Waltz 1981). Despite the incantations of national defence by Abe and Obama, so-called security realists often fail to consider scale and longevity of damage rendered on ecological systems from a military exchange involving the dispersal of radiation contamination, which include human habitats and societies whether they are democratic or otherwise (Helfland 2013).

The Fukushima nuclear disaster has demonstrated that ejected radioactive contaminants do not respect national borders or demarcated zones. This is consistent with a precedent of environmental disasters including Santa Susana, Chalk River, Windscale/Sellafield, Three Mile Island, Semipalatinsk and Chernobyl. As testimonies from local witnesses of radiation-contaminated humans, animals, insects and plants are confirmed by scientific studies, it is increasingly apparent that radioactive contaminants released into the air and the ocean on a daily basis since March 2011 have and will continue to cause harmful effects on multiple ecologies in the near future, in eastern Honshu and in the northern Pacific Ocean and its surrounding coastlines.

Although President Obama declared in Prague in April 2009 that the US would seek 'peace and security in a world without nuclear weapons', the US has not committed to making illegal the production of such weapons. It has not sought a multinational framework for preventing nuclear proliferation and a 'nuclear-weapon-free zone' (NWFZ) in Northeast Asia. Nor has it created its appropriate conditions, by seceding from military drills in the area and incentivizing North Korea to desist from its nuclear weapons development. To the contrary, a legacy of punitive US sanctions and US military (and nuclear) threats along North Korean and Chinese borders (Cumings, 4 November 2013) has only encouraged North Korean intense defensiveness. US military doctrines, ranging from massive retaliation to full-spectrum dominance and pre-emptive global strikes, and US hypocrisy on the CTBT, NPT and nuclear disarmament (Committee on International Security and Arms Control, NAS, 2002, 1–5, 19–34)[xiv], have only encouraged such paranoia.

The radical politics of the Abe government, through risky neoliberal economics, enhanced executive powers and surveillance, and extended transnational military and economic partnerships (the proposed TPP and Japan-EU free trade agreements), reinforces the nexus between preferred big corporations, governments and military and research institutions. In positioning

Japan to channel vitality from the growth hubs of Southeast Asia and India, this aggressive agglomeration is forcing China and Russia toward a closer bloc on the Eurasian landmass. In this situation, the 'clear and present danger' is the sovereign power structure driven the forceful opening of markets and fuelled by nuclear technology.

Given the irreparable toxic burden from nuclear materials that permeate the living environment at the molecular level and which last for roughly 100,000 years, it is unconscionable to continue to prioritize the nuclear industry. Phasing out nuclear technologies and redirecting investment toward the rapidly advancing field of renewable energy production is the only option.

The fact is, entrenching deep rifts between China and Japan through a politics of fear destabilizes the region and legitimates Japan–US militarization. Japan's planned nuclear deployment, interception and retaliatory capacities are as irrational as nuclear deterrence. It only compounds the inertia in responding to the real and present disaster unfolding from Fukushima Daiichi. The ability to demand action to stem the serious contamination of the ocean will be an important test of human intelligence.

Notes

i Following the initial disaster, 'safety' was simply redefined by raising the annual radiation limit by 20 times (from 1 to 20 mSv/y) for civilians, 5 times for workers (from 20 to 100 mSv/y) and 2.5 times for emergency workers (from 100 to 250 mSv/y). Disposal and incineration of radioactive materials have been 'shared' around the nation's territories, while the toxicity of contaminated materials leaking from the plant was under-estimated. In January 2014, TEPCO released much higher measurements for Strontium 90 than was previously estimated emitting from the groundwater surrounding the plant. (See 'TEPCO withheld Fukushima radioactive water measurements for 6 months', *Asahi Shinbun*, 9 January 2014, http://ajw.asahi.com/article/0311disaster/fukushima/AJ201401090060). In contrast to the 178 people TEPCO previously reported to the WHO, TEPCO also admitted that roughly 2,000 workers received doses greater than 100 mSv/y to their thyroid. See 'Fukushima Daiichi Genshiryoku Hatsudensho sagyōsha no naibu hibaku senryo shūsei ni tomonau choki kanri taishōsha no minaoshi to foro-appu ni tsuite', *TEPCO*, 12 July 2012, http://www.tepco.co.jp/nu/fukushima-np/handouts/2013/images/handouts_130712_03-j.pdf; and '2,000 Fukushima workers had thyroid exposure over 100 mSv, TEPCO reported WHO only '178 workers'', *Fukushima Diary*, 20 July 2013, http://fukushima-diary.com/2013/07/2000-fukushima-workers-had-thyroid-exposure-over-100-msv-tepco-reported-who-only-178-workers/.

ii A partial quote from Abe's statement is "Fukushima no jōtai ha kichinto kanri sareteimasu. Keshite Tokyo ni dame-ji wo ataenai.... Kenkō mondai ha ima made mo genzai mo shōrai mo mattaku mondai nai to yakusoku suru / I promise that the condition in Fukushima is strictly under control. In no way will Tokyo receive damage…. The health problem has not been, is not presently and will be not be a problem in future." 'Morning Bird', *Asahi TV*, 8.10 am, 7 September 2013. See also, '20 nen Gorin: IOC Sōkai Purezen Shushō no Hatsugen Yōshi', *Mainichi Shinbun*, 8 September 2013, http://mainichi.jp/sports/news/20130908k0000m050093000c.html.

iii The former PM Mori, for example, threatened that the Olympics would not be possible without nuclear energy production over the next six years. ''Genpatsu Zero' nara gorin henjō shika nai…Mori moto shushō,' *Yomiuri Shinbun*, 18 January 2014. Retrieved on 19 January 2014, from: http://www.yomiuri.co.jp/olympic/2020/politics/20140118-OYT1T00775.htm.

iv For more on nosebleed reports see, Adam Broinowski, 'Fukushima: Life and the Transnationality of Radioactive Contamination', *Asia-Pacific Journal*, 14 October 2013, www.japanfocus.org/-Adam-Broinowski/4009, fn. 30.

v Article 9 of the postwar constitution reads as follows: 'Aspiring sincerely to an international peace based on justice and order, the Japanese people forever renounce war as a sovereign right of the nation and the threat or use of force as means of settling international disputes. In order to accomplish the aim of the preceding paragraph, land, sea, and air forces, as well as other war potential, will never be maintained. The right of belligerency of the state will not be recognized.'

vi 'The emperor shall be the 'head of state' (*genshū*/ 元首) and the symbol of the unity of the people, deriving his position from the will of the people with whom resides sovereign power.' Article 1 Chapter 1 of the LDP draft of an 'autonomous constitution' (*jishu kenpō*) unveiled on April 27 2012, to coincide with the sixtieth anniversary of the coming into force of the San Francisco Peace Treaty in 1951. (See Bix, April 28, 2014).

vii Yasukuni Shintō Shrine houses spirits of Japan's 2.4 million war-dead, and 14 wartime leaders who were found guilty at the Tokyo war crimes tribunal. It houses the Yūshūkan Museum which projects a narrative which supports Japan's wartime aggression in Asia. Against the advice of US Assistant Secretary of State Richard Armitage and Vice President Joe Biden, and on 22 April 2014, Abe made a personal offering to the Chinreisha shrine at Yasukuni. One day before President Obama visited Japan, 150 lawmakers including Internal Affairs Minister Yoshitaka Shindo then attended the Shrine. (See Lies, 22 April 2014).

viii The first NSC meeting comprised Prime Minister Abe, Chief Cabinet Secretary Suga Yoshihide, Defense Minister Onodera Itsunori, Foreign Minister Kishida Fumio, and Deputy Prime Minister Aso Taro in attendance. Watanabe Tsuneo, a close colleague of Nakasone Yasuhiro, and chairman of the Yomiuri Group, was appointed as chairman of the 'Advisory Committee of Intelligence Security'. (See Repeta, March 10, 2014).

ix The Philippines, under a new defence pact, agreed in 2014 to re-open its bases to the US military which it had ousted in 1991 after having occupied them since 1898.

x The Reuters article in Japanese included comments by Obama omitted from the English article, in which he cited Japan and the East China Sea, the Philippines and others in the South China Sea, and Crimea stopped by Russia from joining southern Ukraine. Together with Abe, Obama emphasised the rule of international law and opposed the use of force to change the status quo. (See Reuters, 'Obama Daitōryō ga tokkyaku wa Anpo Jōyaku no taishō to meishin, Chūgoku ni mo hairyo', 24 April 2014, http://jp.reuters.com/article/topNews/idJPKBN0DA07H20140424.)

xi Project for the New American Century (PNAC) is a US neo-conservative think tank of foreign policy specialists and spokespeople which formed in 1997. Having found new influence in the Obama administration, representatives such as National Security Adviser Susan Rice and UN Ambassador Samantha Power have used liberal interventionism to justify invading other nations. Following on from the failure of the US sponsored 'Orange Revolution' in 2004–5, the latest coup was advocated by Victoria Nuland, Secretary of European and Eurasian Affairs, and member of the PNAC.

xii This sort of strategic logic can be found in works such as Nakanishi Terumasa (ed.), 'Nihon Kaku Busō' no Ronten, Tokyo: PHP Kenkyūjo, 2006.

xiii The USS Midway and Kitty Hawk, both of which carried nuclear weapons, were stationed at Yokosuka for two decades from 1972. (See Kristensen, 1999). For the secret agreements between the US and Japan, see Wampler, 13 October 2009. For the debate under Satō see Kase, 2001.

xiv The National Academy of Sciences (US) argued that the Comprehensive Nuclear-Test-Ban Treaty (CTBT) would permit the United States to maintain technical reliability of its nuclear weapons by means other than nuclear testing that would be consistent with the U.S. Stockpile Stewardship Program that was established in 1994. See Committee on Technical Issues Related Ratification of the Comprehensive Nuclear Test Ban Treaty, 2002.

References

Giorgio Agamben, *State of Exception*, Chicago, University of Chicago, 2005.

Akiyama Nobumasa, 'The Socio-Political Roots of Japan's Non-Nuclear Posture,' in Benjamin L. Self and Jeffrey W. Thompson (eds.), *Japan's Nuclear Option: Security, Politics, and Policy in the 21st Century*, Washington, DC: Henry L. Stimson Center, 2003.

Mizuho Aoki, 'Ukraine crisis has Japan on horns of a dilemma Stance against territory violations endangers closer ties with Russia', *The Japan Times*, 30 April 2014.

Associated Press, 'Navy to Deploy Aegis Destroyer in Japan,' *Japan Times*, 17 November 2006.

Jeremy Bender, 'Why Japan's smaller military could hold its own against China,' *Business Insider*, 6 May 2014, http://www.businessinsider.co.id/japans-smaller-military-could-match-china-2014-5/#.U4FLLS-sPPl.

Walter Benjamin, *Critique of Violence, Selected Writings; Vol 1; 1913–1926*. Marcus Bullock and Michael Jennings (eds.), London, The Belknap Press of Harvard University Press, 1996.

Herbert P. Bix, 'Abe Shinzo and the U.S.-Japan Relationship in a Global Context', *The Asia-Pacific Journal*, Vol. 12, Issue 17, No. 1, April 28, 2014.

Hans Blix, 'Israel possesses nuclear weapons: Former IAEA head,' *Press TV*, 12 May 2014, http://www.presstv.ir/detail/2014/05/12/362349/israel-possesses-nuclear-weapons/; 'International nuclear body to curry favor with Israel: Exposed', *Press TV, 15 April 2014*, http://www.presstv.ir/detail/2014/04/15/358676/nuclear-body-to-curry-favor-with-israel/.

Adam Broinowski, 'Fukushima: Life and the Transnationality of Radioactive Contamination,' The Asia-Pacific Journal, Vol. 11, Issue 41, No. 3, October 14, 2013.

Subhomoy Bhattacharjee, Anil Sasi, 'Japan wants slice of nuclear pie, warms up to liability law', *Indian Express*, 21 June 2014, http://indianexpress.com/article/business/business-others/japan-wants-slice-of-nuclear-pie-warms-up-to-liability-law/, .

Kurt M. Campbell and Tsuyoshi Sunohara, 'Japan: Thinking the Unthinkable', in Campbell, Robert J. Einhorn, and Mitchell B. Reiss (eds.), *The Nuclear Tipping Point: Why States Reconsider Their Nuclear Choices*, Washington, D.C.: Brookings, 2004.

Brahma Chellany, 'Asia's emerging democratic axis', *Japan Times*, 31 January 2014.

Committee on Technical Issues Related Ratification of the Comprehensive Nuclear Test Ban Treaty, Committee on International Security and Arms Control, National Academy of Sciences, *Technical Issues Related to the Comprehensive Nuclear Test Ban Treaty*, Washington, National Academies Press, 2002.

Bruce Cumings, 'Korea: Forgotten Nuclear Threats,' *Le Monde Diplomatique*, December 2004; Foster Klug, *Huffington Post*, 4 November 2013, http://www.huffingtonpost.com/2013/04/11/u-s-south-korea-military-drills-north-korea_n_3060577.html.

Chester Dawson, 'Japan shows off its missile defence system,' *Wall Street Journal*, 9 December 2012.

Defense Industry Daily, 'Japan's next F-X fighters: F-35s win round 1, 25 August 2013, *Defence Industry Daily*, http://www.defenseindustrydaily.com/f22-raptors-to-japan-01909/.

Ken Jimbō, 'Rethinking Japanese Security: New Concepts in Deterrence and Defense,' in Thompson and Self (eds.), *Japan's Nuclear Option*, pp. 33, 41–4.P. A. Dionisopoulos, 'The No-War Clause in the Japanese Constitution', *Indiana Law Journal*, Vol 31, No 4, 1956.

John Foster Dulles (5 September 1951), "San Francisco Peace Conference speech", Database of Japanese Politics and International Relations, Tokyo, University of Tokyo.

Editor, Dianuke, 'India, Japan to strengthen ties with economic, Shinzo Abe's nuclear marketing trip to India not a success: *Resisting Abe's Sales Pitch,*' *Dianuke, 5 February 2014,* http://www.dianuke.org/resisting-abes-sales-pitch/#sthash.ug3w4MvO.dpuf.

Editor, Korea Joongang Daily, 'Japan's gov't revisits Kono evidence', *Korea Joongang Daily*, http://koreajoongangdaily.joins.com/news/article/Article.aspx?aid=2989450.

Editor, Kyodo, 'Japan Can Legally Be Allowed to Possess Nuclear Arms, Government Says,' *Kyodo News*, 14 November 2006.

Editor, Japan Times, 'Exports that defy reason', *Japan Times*, 20 April 2014, http://www.japantimes.co.jp/opinion/2014/04/20/editorials/exports-that-defy-reason/#.U4J0JC-sPPk.

Editor, Daily Yomiuri, 'Government May Ask U.S. for Details of Nuclear Defense Strategy,' *Daily Yomiuri*, 21 March 2007.

Editor, Japan Times, 'Nakasone Proposes Japan Consider Nuclear Weapons,' *Japan Times*, 6 September 2006.

Editor, Japan Times, 'Sasebo May Host Site for Maintaining Interceptor Missiles,' *Japan Times*, 5 December, 2006.

Editor, Reuters, 'Obama Daitōryō ga tokkyaku wa Anpo Jōyaku no taishō to meishin, Chūgoku ni mo hairyo', 24 April 2014, http://jp.reuters.com/article/topNews/idJPKBN0DA07H20140424.

John E. Endicott, 'The 1975–76 Debate Over Ratification of the NPT in Japan,' *Asian Survey* 17, March 1977.

Michel Foucault, *History of Sexuality*, vol. 1, London, Penguin, 1990.

David Frum, 'Mutually Assured Disruption,' *New York Times*, 10 October 2006, p. A25.

Furukawa Katsuhisa, 'Nuclear Option, Arms Control, and Extended Deterrence: In Search of a New Framework for Japan's Nuclear Policy,' in Self and Thompson (eds.), *Japan's Nuclear Option*, 2003.

Francis Gavin, 'Blasts from the Past: Proliferation Lessons from the 1960s', *International Security*, Vol. 29, No. 3, Winter 2004/5.

Nancy Gibbs, Hannah Beech, 'The Patriot: Shinzo Abe speaks to Time', *Time*, 17 April 2014, www.time.com/65673/shinzo-abe-japan-interview/.

Mel Gurtov, 'Back to the Cold War? The US-China Military Competition', March 10, 2014, http://www.chinausfocus.com/foreign-policy/back-to-the-cold-war-the-us-china-military-competition/.

Hara Takuji, 'Social Shaping of Nuclear Energy: Before and after the Disaster', R. Hindmarsh ed., *Nuclear Disaster at Fukushima Daiichi: Political, Social and Environmental Issues*, UK, Routledge, 2013.

David Harvey, *The New Imperialism*, Oxford, Oxford University Press, 2003.

Ira Helfland, 'Nuclear Famine: Two Billion People at Risk: Global Impacts of Limited Nuclear War on Agriculture, Food Supplies and Human Nutrition', *International Physicians for the Prevention of Nuclear War*, *Physicians for Social Responsibility*, November 2013.

Seymour Hersh, 'The red line and the rat-line', *London Review of Books*, 17 April 2014, vol. 36, no. 8, pp. 21–4.

Hisane Masaki, 'Japan Joins the Race for Uranium Amid Global Expansion of Nuclear Power,'' *Japan Focus*, April 22, 2006, http://japanfocus.org/-hisane-masaki/1626.

The Institute for International Policy Studies (IIPS), 'A Vision of Japan in the 21st Century,' September 5, 2006, www.iips.org/National%20Vision.pdf.

Japan and United States Governments, 'Joint Statement: The Japan-US Alliance of the New Century', Prime Minister Junichirō Koizumi, President George W. Bush, 29 June 2006.

JAIF, 'Japan's Upper House Approves Four Nuclear Cooperation Agreements: Foreign Minister Pursues Cooperation after Fukushima Accident,' *Japan Atomic Industrial Forum*, 19 December 2011, http://www.jaif.or.jp/english/news_images/pdf/ENGNEWS01_1324278528P.pdf

Japan Defense Agency, *Defense of Japan 2006*, Tokyo, Japan Defense Agency, 2006

Jiji, 'Outrage derails manga series 'Oishinbo' for Fukushima nuclear crisis depiction', *Jiji*, 18 May 2014.

Eric Johnston, 'Rokkasho Drawing Proliferation Flak: Multinational Controls Urged to Deter Nuclear Arms-Seeking Copycats,' *Japan Times*, November 15, 2005.

Eric Johnston, 'Rokkasho Tests Break Plutonium Pledge, Activists Tell IAEA,' *Japan Times*, January 6, 2006.

Eric Johnston, 'Going nuclear: How close has Japan come? We examine the historical debate on the country's atomic ambitions', *Japan Times*, 10 May 2014.

Matake Kamiya, 'Nuclear Japan: Oxymoron or Coming Soon?' *Washington Quarterly* 26, Winter 2002/03, pp. 65–7

Yuri Kase, 'The Costs and Benefits of Japan's Nuclearization: An Insight into the 1968/70 Internal Report,' *Nonproliferation Review* 8, Summer 2001, pp. 58–9.

Naomi Klein, The Shock Doctrine: The Rise of Disaster Capitalism, Melbourne, Penguin, 2007.

Charles Krauthammer, 'The Japan Card', *Washington Post*, 3 January 2003.

Hans Kristensen, 'Japan under the Nuclear Umbrella', *Nautilus Strategic Research Institute*, 1999, http://oldsite.nautilus.org/archives/library/security/papers/Nuclear-Umbrella-4.html.

Kyodo News, 'Nukes Still Not Welcome: Cabinet,' *Japan Times*, 21 December 2006.

David Lague, 'Chinese military's secret to success: European engineering', *Reuters*, 20 December 2013, http://in.reuters.com/article/2013/12/19/breakout-submarines-special-report-pix-g-idINL4N0JJ0FM20131219.

John J. Mearsheimer, *The Tragedy of Great Power Politics*, New York: W.W. Norton, 2001.

Kyle Mizokami, 'The Chinese military is a paper dragon', *War is Boring*, https://medium.com/war-is-boring/8a12e8ef7edc.

Elaine Lies, 'Japanese minister, MPs visit Yasukuni Shrine on eve of Obama visit', *Reuters*, 22 April 2014.

Mike Mochizuki, 'Japan tests the Nuclear taboo', *Non-Proliferation Review*, vol 14, No 2, July 2007.

Nakanishi Terumasa, 'Nuclear Weapons for Japan,' *Japan Echo* 30, October 2003.

Nakanishi Terumasa (ed.), *'Nihon Kaku Busō' no Ronten*, Tokyo: PHP Kenkyūjo, 2006.

Nakanishi Terumasa, Nishioka Chikara, 'Anpō kara Hajimaru 'Sengo kara no Dakkyaku' to Nihon no Kaku Busō', *Seiron*, December 2006, pp. 60–1.

Hiroko Nakata, 'Japan Makes It Official: More Punitive Steps Kick In,' *Japan Times*, October 14, 2006

Masashi Nishihara, 'North Korea's Trojan Horse,' *Washington Post*, 14 August 2003.

Office of Intelligence Research, 'The outlook for nuclear weapons production in Japan', *Division of Research for Far East*, US State Department, Washington, 2 August 1957.

Okazaki Hisahiko, 'Time to Change Our National Security Strategy,' *Daily Yomiuri*, 23 April 2006.

Okuyama Toshihiro, 'U.S. alarmed about plutonium stockpile growing from Rokkasho plant', *Asahi Shinbun*, 13 April 2014.

Tomohiro Osaki, 'Depictions of characters' radiation-induced illnesses 'give voice' to silent Fukushima residents. 'Oishinbo' editor defends manga', *Japan Times*, 19 May 2014.

William C. Potter, 'India and the New Look of U.S. Nonproliferation Policy,' *Nonproliferation Review* 12, July 2005.

A.V. Ramana, 'The Power of Promise: Examining Nuclear Energy in India,' Penguin Viking, 2012.

Austin Ramzy, 'China complains about Plutonium in Japan', *New York Times*, 10 June 2014, http://sinosphere.blogs.nytimes.com/2014/06/10/china-complains-about-plutonium-in-japan/?_php=true&_type=blogs&_php=true&_type=blogs&_php=true&_type=blogs&_r=2.

John Reed, 'Surrounded: How the U.S. Is Encircling China with Military Bases,' *Foreign Policy*, 20 August 2013.

Lawrence Repeta, 'Japan's 2013 State Secrecy Act—The Abe Administration's Threat to News Reporting,' *The Asia-Pacific Journal*, Vol. 12, Issue 10, No. 1, March 10, 2014.

Mari Saito, Anthony Slodkowski, 'Japan's Homeless Recruited for Murky Fukushima Clean-up', *Reuters*, 30 December 2013, http://www.reuters.com/article/2013/12/30/us-fukushima-workers-idUSBRE9BT00520131230.

Richard Samuels, *The Business of the Japanese State: Energy Markets in Historical Perspective*, Ithaca, Cornell University Press, 1987.

Michael Schaller, 'Japan and the Cold War, 1960–1991', *The Cambridge History of the Cold War*, Cambridge, Cambridge University Press, 2014.

Carl Schmitt, Political Theology: Four Chapters on the Concept of Sovereignty, Chicago, University of Chicago Press, 2006.

Sheila A. Smith, 'North Korea in Japan's Strategic Thinking, *The Asan Forum*, 7 October 2013, http://www.theasanforum.org/north-korea-in-japans-strategic-thinking/.

Yoshihide Soeya, 'Japan: Normative Constraints Versus Structural Imperatives,' in Muthiah Alagappa, (ed.), *Asian Security Practice: Material and Ideational Influences*, Stanford, CA, Stanford University Press, 1998.

Suzuki Manami, *Kaku tai Kokuka suru Nihon: Heiwa Riyō to Kaku Busō Ron* (A Japan Moving Toward a Nuclear Great Power: Peaceful Use and the Argument for Nuclear Armament), Tokyo, Heibon sha, 2006.

Hiroko Tabuchi with Makiko Inoue, Matthew Wald and David Sanger, 'Japan Pushes Plan to Stockpile Plutonium, Despite Proliferation Risks', 9 April 2014, http://mobile.nytimes.com/2014/04/10/world/asia/japan-pushes-plan-to-stockpile-plutonium-despite-proliferation-risks.html.

Ashley J. Tellis, 'India as a New Global Power: An Action Agenda for the United States,' Washington, DC, Carnegie Endowment for International Peace, 2005.

Ida Torres, 'Japan launches National Security Council, discusses China and North Korea in first meeting', 5 Dec 2013.

Tsuchida Atsushi, 'The Nuclear Arming of Japan', *Japan Society of Physics*, Paper presentation, Tokyo, 1 August 1993.

Christopher P. Twomey, 'The Dangers of Overreaching: International Relations Theory, the U.S.-Japan Alliance, and China,' in Self and Thompson, (eds.), *An Alliance for Engagement: Building Cooperation in Security Relations with China*, Washington, 2002.

Kenneth Waltz, 'The spread of nuclear weapons: More may be better', *Adelphi Papers*, Number 171, London: International Institute for Strategic Studies, 1981.

Robert Wampler, 'Nuclear Noh Drama: Tokyo, Washington and the case of the missing agreements', *The National Security Archive*, 13 October 2009.

Timothy Weiner, *Legacy of Ashes: The History of the CIA,* New York, Doubleday, 2007.

Yamada Takao, 'Nuclear armed Japan is not out of the question', *Mainichi Shinbun*, 25 March 2012.

Mari Yamaguchi, 'Defense Chief Says Nuclear Passage Through Japan Waters May Be Unavoidable,' *Associated Press*, 26 November 2006.

Reiji Yoshida and Masami Ito, 'Japan Stands Firm with Sanctions on North Korea', *Japan Times*, 2 November, 2006.

The Political Challenge of Denuclearizing Japan

Richard Wilcox and Tony Boys

Our dysfunctional global political system is paving the way to our Nuclear Armageddon. OK, it is not "our" political system, it's "theirs," and it functions perfectly well for the top 0.01 percent of oligarchs who really run things. But even the filthy rich have children and grandchildren whose DNA is likely to be harmed by the Radioactive Environmental Armageddon humanity is leaping toward. Nominally, we the people are represented by our governments, but this is not always the case. Governments are arguably Corporations and Corporations are run by Banks, their Sinister Financial Formulae putting Profits before People ("Walter Burien," 2012).

If we had democracy in Japan or elsewhere there could be no doubt that public pressure since the 3/11 nuclear disaster would have already forced the phase out of nuclear power and a program of renewable energy to be quickly adopted. Our problems are not technical, but political, and relate to a maldistribution of power and wealth in the world. The existing economic system does not prioritize human welfare and environmental preservation, but Ponzi scam financial swindles based on derivatives, high frequency computerized trading and naked short selling of stocks. The new movie "The Wolf of Wall Street" described as "nauseating, pornographic and soul-crushing" tells it all (Duke, 2014).

Things are getting worse, much worse and fast. Anti-nuclear hero Harvey Wasserman coined the slogan "No Nukes" after the Three Mile Island nuclear disaster in the US in the 1980s. Wasserman's encyclopedic knowledge of nuclear coverups and criminality displayed in a 2014 article (included as a chapter in this volume) is unparalleled, and profoundly disturbing in its detail (Wasserman, 2014). Fine critiques of the dangers of nuclear radiation have also been written by Dr. Chris Busby and Paul Zimmerman (Busby, 2006; Nadesan, 2013; Zimmerman, 2009). The Fukushima nuclear disaster was, in the great scheme of things, just one particularly nasty disaster among a pool of ongoing nuclear disasters ranging from small scale leakages around the world to potential meltdowns in our not so distant future.

Nuclear power is ideal as a technology that is opposed to democratic transparency. Compared to solar power, which average people can learn to operate safely on the rooftops of their homes, nuclear power is incompatible with democracy and a potential source of the ultimate state terrorist weapon, a nuclear weapon. Nuclear power plants are sitting targets for secret drone strikes which make them a potential dirty bomb. In addition, the complexity of running a nuclear plant with all the engineering and safety requirements versus other forms of energy production (i.e. a campfire) is mind boggling (Wilcox, 2013). The potential danger of a plane or drone's computer system being hacked by terrorists or malicious governments and then being deliberately crashed by remote control into a reactor or spent fuel pool at a nuclear power plant is confirmed by ex-CIA agent Chip Tatum ("Gene 'Chip' Tatum," 2014). It is also now well known that nuclear power plant operating systems are vulnerable to cyberattack by Stuxnet (e.g. "Stuxnet worm").

Wasserman points out that the reason we don't have safe and renewable forms of energy today is because the power companies and the shadowy interests behind them enjoy a monopoly on electrical, and by extension, economic and political power. The psychopathic nuclear oligarchy is happy to make a lot of money at the expense of the planet. The social system based on economic incentives and political favors controlled by the nuclear industry has created a veritable "iron nuclear triangle" between the nuclear industry, the bureaucrat/politician nexus and business interests in Japan (Kingston, 2012), virtually assuring that progressive change is impossible within the current paradigm (Meyer, 2011).

That is why one of the authors, who has been part of the anti-nuclear movement and an energy researcher in Japan for many years, has maintained (however unrealistic it may be) that there's only one way to stop nuclear power in Japan - massive civil disobedience. But since people are happy to elect conventional, liberal and economically growth-oriented politicians like Abe and Masuzoe (see below), then the populace are naive or self-destructive participants who, while suffering the consequences of their choices, ruin the chances of a radiation-free environment and freedom of choice of energy supply systems for the rest of us.

One of the main reasons Prime Minister (PM) Naoto Kan, who was in office at the time of the Tohoku Earthquake Disaster on March 11, 2011, was forced from office was due to pressure from the nuclear lobby specifically because of Kan's strong anti-nuclear/pro-renewable energy stance (Tabuchi, 2011).

Just over a year after the Fukushima nuclear disaster had occurred, a petition signed by 320,000 Tokyoites was submitted calling for a referendum on the future of nuclear power plants ("Tokyo assembly," 2012). The fact that such a huge number of people signed it clearly indicated that if a referendum were to have been held, the citizenry would have voted to phase out nuclear power, as has happened in German and Italian referendums.

However, "[a]ssembly members voted 2-1 ... to reject a draft ordinance calling for the referendum, which had been forced onto the agenda by the signature petition from the Tokyo public." This is clear proof that elected representatives in Japan do not represent the will of their constituencies on important issues.

> 'I feel frustrated and empty,' said Saori Kano, 45, a homemaker
> from Tokyo's Setagaya Ward who helped collect signatures in

the Tokyo drive. 'My goal is to change the nuclear power policies that have been left to the central government to decide.'

Ms. Kano very neatly voices the frustration of large numbers of ordinary people, who feel that they have been blatantly dispossessed of their democratic rights when it comes to nuclear policy.

Three Years of Fukushima Radiation

As Japan's Kobe University seismologist, Katsuhiko Ishibashi, noted to a government panel in 2005, "[a]n earthquake and its seismic thrust can hit multiple parts" of a nuclear plant and result in a "severe accident" (Hongo, 2011). No kidding! At the time of Ishibashi's testimony to a government panel (ironically his name means "stone bridge" in Japanese), monied interests in heavy industry pressured the government to ignore his advice regarding nuclear safety standards.

And what of the people whose lives have been shattered and displaced by the triple meltdowns that occurred in Fukushima in March 2011? Tens of thousands of people have been forgotten, cast aside, blatantly dispossessed with inadequate compensation, as if they were just disposable human garbage (Gundersen, 2014, February 3; Wilcox, 2014, April 12).

Further indication that Japan's political system is undemocratic is the level of political denial regarding the ongoing nuclear crisis that remains technically unresolved to date (Obayashi, 2014). For example, the reactor 3 fuel pool will literally take decades to clean up, if not a nearly impossible situation to remedy (Gundersen, 2014, February 13).

From the outset of Japan's nuclear program, which was opposed by many, up to the present day, transparency of operations of the nuclear industry has been woefully lacking ("Nuke Info," 2014). After World War II the Japan Scientists Council recommended that the government not adopt nuclear power because of the obvious dangers, but the "peaceful atom" propaganda of the day prevailed and nuke plants were rapidly put in place, whether on top of, or near, geological fault-lines (Cohen & McKillop, pp. 129 - 130, 2012).

This harkens to our point that nuclear technology is inherently opposed to democracy. Symbolically, in the face of the horrendous catastrophe that has unfolded as the three reactors at Fukushima No. 1 Nuclear Power Station melted down, the leaders of the successive Democratic Party of Japan (DPJ) and Liberal Democratic Party (LDP), along with the bureaucrats in the Ministry of Economy, Trade and Industry, have consistently attempted to deceive the public that there is no danger and that matters are "under control." Any real democracy would have tarred, feathered, and thrown these scoundrels out on their ears.

Note that when former Tokyo governor, Mr. Inose, publicly contradicted statements made by Prime Minister Abe, Inose suddenly found himself embroiled in a "scandal" and removed from office ("Tokyo governor," 2013). Abe fibbed to the Olympic Committee that the Fukushima nuclear crisis was "under control," and Inose naively refuted Abe's obvious lies, thus causing embarrassment to Prime Malefactor Abe. This precedent was not allowed to stand, especially given Abe's plans to kickstart Japan's nuclear program:

> Speaking at a press conference on Friday, Inose refuted Abe's claim, telling reporters that the water leaks at the plant were 'not necessarily under control,' Fuji TV reported Saturday. Hours before the IOC declared Tokyo the host city for the 2020 Games, Abe flew to Buenos Aries from the G20 summit in St Petersburg, Russia, to give an emphatic speech in English declaring that radiation from the leakage would not impact waters outside the immediate vicinity of the plant. Inose said, 'The government must acknowledge this as a national problem so that we can head toward a real solution.'

According to the Japan Communist Party (JCP) newspaper, ruling LDP candidates are threatened by the electric power companies with removal of campaign support unless they promote nuclear power ("Electric companies," 2014). LDP politicians are sock puppets who say what the nuclear industry wants them to. Early in January 2014, when a survey on Japan's energy strategy was circulated to LDP parliamentary members, the Federation of Electric Power Companies of Japan (among other things responsible for nuclear propaganda on behalf of the ten power companies in Japan that own nuclear reactors) sent sheets containing pro-nuclear "model answers" to the Diet-men, some of whom no doubt used the models rather than bother to think about what to write ("Blatant behind," 2014).

But why does the conventional political power structure (consisting largely of the LPD and Komeito) in Japan want nuclear power so much that they are willing to lie, subvert the democratic process and endanger the current population (including themselves, but with little likelihood they will ever be "compensated") and future generations (including their own descendents)?

1) Money/power and livelihoods for the inhabitants of the "nuclear village." The term nuclear village (*gensihryoku mura*) was coined by Iida Tetsunari, an anti-nuclear and pro-renewable energy activist ("nuclear village"). The idea of a "village" having cozy relations within its membership, is employed in an ironic and cynical sense here, meaning the powerful interests and players involved have a cozy relationship but at the expense of the rest of the society. The terms "lobby," "cartel" or "nexus" could be used interchangeably to refer to the collective of power companies, nuclear plant makers, nuclear plant service companies, academics and researchers in the nuclear field and everyone that has a direct or indirect interest in nuclear power. In addition, politicians who receive donations and the media who receive advertising revenues and program sponsorships are all happy members of the nuclear village (Meyer, 2011);

2) For the sake of the Japanese economy, since fossil fuels are now expensive and their import for thermal power generation is dragging Japan's trade

balance into deep red ink (Mogi & Ujikane, 2014, Boys, 2000, p.26). Business circles, led by the Keidanren Chairman Masahiro Yonekura, are in the forefront of the push to have as many nuclear power plants restarted at the earliest possible date ("Japan's U-turn," 2012).

3) The US (and to some extent UK and France) do not wish to see Japan abandon civil nuclear power because they do not wish to see a hole punctured in the notion of the "peaceful atom" ideology (Favole & Tennille, 2011). Japan's threat to abandon nuclear power for safety reasons frightens countries which hold nuclear weapons capabilities since it robs them of their only rational ground for maintaining the production of fissile materials. Without this, they fear, overwhelming public pressure to be finally rid of civil and military nuclear applications would force power holders to relinquish the very bedrock of their military power – their nuclear arsenal. Thus Japan, with its civil nuclear capability but apparent non-military capability, acts as a keystone maintaining the credibility, and thus the unshakeability, of the mythic "peaceful atom" structure. Ironically, it is an open secret that Japan possesses quickly achievable nuclear weapons status (Trento, 2012).

For these reasons it is simply "impossible" for Japan to abandon nuclear power, despite the role it plays in the ongoing destruction of the physical nation from radioactive contaminants.

The Tokyo Election Circus: Nuclear Power Wins - People Lose

Come one, come all to witness the clowns, acrobatic rhetoric and genetically mutated freak show of all form of chimera, creature and species, in the battle for the National Soul – the Tokyo Metropolis gubernatorial election of February 9, 2014. What made this election important for the nuclear village and the LDP was that it was just possible that a well known and respected lawyer (Kenji Utsunomiya) with strong anti-nuclear credentials might win. For this reason, PM Abe was reported to have expressed concern that nuclear issues might become the dominant focus of the election (Tokyo gubernatorial election, 2014). A politically anti-nuclear Tokyo, inhabited by over 13 million people, who have little option at present but to buy TEPCO's electricity, made it imperative that this nightmare outcome be avoided at all costs. Thus it was necessary to ensure that electoral debates focused on other, more local, issues and that the, or any, anti-nuke candidate did not win.

The morning after the 2014 gubernatorial election in Tokyo, *The Japan Times* headline screamed "Masuzoe scores landslide victory in Tokyo gubernatorial race" ("Masuzoe scores," 2014): "It was more than the combined votes for Utsunomiya, 67, former head of the Japan Federation of Bar Associations, who was a distant second and Hosokawa, 76, who came third." The election provided a new governor who is an Abe stalwart ready to lend support to his militaristic political agenda and plans to restart idle nuclear reactors.

What's going on? Morihiro Hosokawa was Japanese Prime Minister (PM) from August 1993 to April 1994. He was assisted in his gubernatorial bid by Junichiro Koizumi, PM from April 2001 to September 2006. This is very suggestive, given that Koizumi was the immediate predecessor to Shinzo Abe, the current PM, in his first term as PM from September 2006 to September 2007. Koizumi also held a cabinet position in Abe's first term of office. The very fact that Hosokawa decided to run, and that Koizumi spared no effort to aid his campaign, almost to the extent that it sometimes looked as if they were running on some kind of joint candidacy, reeks of stale cigar smoke and backroom deals of an old boys' club.

After Naoki Inose resigned as governor on December 24, 2013, Utsunomiya, who had come second in the gubernatorial race with Inose in 2012, was the first to announce his candidacy, on December 28. ("Utsunomiya 1st") Masuzoe officially announced his candidacy on January 14, 2014, after picking up the support of the LDP delegation in the Tokyo metropolitan assembly. Hosokawa, formerly of the LDP, but later of the DPJ, egged on by the veteran LDP politician Koizumi, finally announced his candidacy on January 22, the final day before official campaigning began (Tokyo gubernatorial election, 2014). All three candidates ran as independents with the support of different, sometimes multiple, political parties, not as official party candidates.

Officially announcing his candidacy on January 14, Masuzoe claimed to be in favor of a nuclear phaseout. The following day, he met with LDP Chief Secretary Shigeru Ishiba, thus confirming LDP support for his candidacy, and thereafter appears to have restrained his statements on energy policy to saying that he would encourage renewable energy and that it was necessary to find a way out of the dependence on imported fossil fuels, which was more the territory of the national government than the Tokyo Metropolis (Masazoe Yoichi).

Utsunomiya called for providing a check against Abe's policies, closure of all Japan's nuclear plants, restrictions on the budget for the 2020 Tokyo Olympics and making Tokyo "a secure city where people can live and work" (Utsunomiya 1st). After the emergence of Hosokawa's candidacy, former Prime Minister Naoto Kan called on Utsunomiya to leave the race out of fears that he would split the anti-nuclear vote in Tokyo (Tokyo gubernatorial election, 2014). Despite Kan's open anti-nuclear stance, this appears suspicious given the fact that Utsunomiya had declared his intention to run very early, while Hosokawa had left it until virtually the last possible moment.

As Hosokawa made his candidacy for governor official, he pledged that he would try to end Tokyo's dependence on nuclear power and would set up a strategic panel of experts to explore basic energy policies for an end to nuclear power. Finally revealing his election pledges after two delays, Hosokawa stated at a news conference on January 22 that "In order to realize a Tokyo that is not dependent on nuclear energy, I would prompt the public and private sectors to generate renewable energy as well as to ask for cooperation from the residents of Tokyo to conserve energy" [Kameda & Yoshida, 2014].

Is it possible the ex prime ministerial Hosokawa/Koizumi (H/K) election campaign team was sent out by the nuclear village/LDP to siphon votes off from the only strong and principled candidate, Kenji Utsunomiya? Although early polls showed Masuzoe's support running at about 40 percent, thus making it likely that he would be elected, his victory was by no means certain. There was still a possibility that Utsunomiya might galvanize anti-nuke sentiment in Tokyo to go on to win, and it was thus necessary to find a means to avoid this outcome. Alternatively, it may be that there was a genuine internal conflict regarding nuclear issues within Japan's ruling hierarchy of Tokyo University graduates, the men who go on to rule the country and become LDP stalwarts. Having become very outspoken against nuclear power after March 11, 2011, Koizumi was totally out of step with the mainstream of the LDP. But why would that translate into persuading Hosokawa, now of the DPJ, to run? Since the LDP had declared support for Masuzoe, Koizumi himself could not very well suddenly decide to join the race, and in fact had declined an invitation to run from Yoshimi Watanabe, the then leader of the small Your Party (Tokyo gubernatorial election, 2014). Perhaps H/Ks' intentions were pure but their approach just turned out to be destructive. Utsunomiya is a Ralph Nader type lawyer who fights for the average people and worker, and is quite obviously and genuinely anti-nuke. It seems that the establishment candidate Hosokawa, with Koizumi (who famously helped to destroy Japan's middle class by "restructuring the economy" in the first half of the 2000s) jumped into the race for the purpose of spoiling Mr. Utsunomiya's chances (Kameda, 2014).

On the other side of the political spectrum to the right wing LPD we have the left wing Japan Communist Party (JCP). The authors of this chapter recognize the unequivocal evil of the role in communism in 20th century history, including mass killings of innocent people and the destruction of untold numbers of cultures according to a problematic Marxist ideology (Solzhenitsyn, 1973). Despite the JCP's authoritarian streaks, it must also be acknowledged that their role in Japanese politics has been to function as one of the few forces to counterbalance the LDP/big business hegemony, and indeed they have often stuck up for the average worker and the downtrodden in society. This is why, although Utsunomiya is essentially an independent, he joined forces with the JCP in the last election.

It is perhaps clear that one of the two anti-nuke candidates could have decided to endorse the other and drop out of the race. If so, we contend that it would have seemed more natural for Hosokawa to be stepping down in favor of Utsunomiya. Why should Hosokawa, with a past record far less obviously anti-nuke, and having just managed to cobble together his candidacy and election pledges at the last moment, take preference over Utsunomiya? If it was name value, then all Hosokawa needed to do was appear with Utsunomiya in public at campaign speeches, perhaps with Koizumi in attendance as well. *But that would* have meant redrawing the whole post-war political map of Japan. Major personalities from the LDP and DPJ (most of whom are former LDP members) are NOT going to be seen endorsing anti-nuke and JCP-supported candidates in an election considered to be crucial by the LDP top brass. Such an unimaginable endorsement would have triggered an explosive media event and could have fired up the Tokyo populace to the extent of sidelining Masuzoe and making it a virtual certainty that Utsunomiya won.

The reverse, Utsunomiya dropping out and endorsing Hosokawa's candidacy, would have been happily acceptable to the LDP, but the JCP-Utsunomiya coalition would have ruined the credibility they have gained through their anti-nuke stance by compromising with politicians cut from the same cloth as the Abe crowd and whose "reborn" anti-nuke credentials are far less robust. It is hard to imagine the hard core and uncompromising JCP ideologues joining hands with such a coalition in the first place, as noted above, and with justifiable skepticism on their part. In the end, Utsunomiya refused to give way to a dubiously anti-nuke candidate, who revealed his election pledges too late even to hold a public debate, and the vote was split.

One hint that Hosokawa/Koizumi were not quite as seriously anti-nuclear as they were trying to make out, and that Hosokawa's participation in the election was nothing more than a cynical stunt, was the overly casual dress style of the two as they stood atop the loudspeaker campaign car to make speeches in the run up to the election. Conventional politicians never fail to appear in public on important occasions in the business uniform of dark suit and tie. But here was Hosokawa garbed in black coat and expansive green muffler (it is considered impolite in Japan to speak to audiences while wearing a muffler) and Koizumi fitted out with a padded greenish-yellow jacket. All of this was plainly visible on the NHK 7 p.m. news in the cold days before the election. The message was also clearly picked up by in a cartoon on page 2 of the Tokyo Shimbun of February 11, when the final official results were published, prominently featuring the duo's unusual fashion sense while Masuzoe and Utsunomiya sported suits and ties. To LDP supporters, it was patently obvious who not to vote for. There was no way they could fail to get the message and vote the "wrong" way. Therefore, the 956,063 votes Hosokawa gathered were due largely to voters who were fooled by the anomalous "green" image and the empty anti-nuclear rhetoric of asking people to become involved in renewables and energy conservation, none of which is necessarily contradictory to NPP restarts.

Another clue of inauthenticity on the part of LDP is that the LDP does have an authentically anti-nuclear Diet member, Taro Kono, whose constituency is Section 15 of Kanagawa Prefecture (which includes Hiratsuka City, where he lives), a very short distance from Tokyo. Mr. Kono with his impeccable anti-nuclear credentials was not to be seen on TV or elsewhere endorsing the anti-nuclear Hosokawa candidacy for the Tokyo governorship. Perhaps he was not terribly impressed by the H/K duo's sincerity.

Utsunomiya received 982,595 votes in the Tokyo election, which is an admirable number given the poor overall voter turnout. Masuzoe's 2,112,979 votes being nearly matched by a total of 1,938,657 votes for Utsunomiya and Hosokawa (only 174,322 short) is evidence of growing dissatisfaction among Tokyoites with the LDP/Komeito coalition ("Utsunomiya vote total," 2014). (Hosokawa's vote total was 956,063, and the total votes cast was 4,930,251 from an electorate of 10,685,343 – a 46.14 percent turnout compared to 62.6 percent in 2012) (Tokyo gubernatorial election, 2014). Had Hosokawa not run, and setting aside the likelihood that foul play, vote fraud or other tactics could have been used against Utsunomiya, a bit more organization in Utsunomiya's camp, for example in persuading younger people to vote, could possibly have seen him win the election. That was how important Hosokawa's electoral intervention was for the nuclear village. The spoiler strategy turned out nicely for Abe.

The low voter turnout (46.14 percent, the third lowest ever) was due to Tokyo's worst blizzard in 45 years happening just the day before the election, not to mention overall apathy and hopelessness among voters. Komeito, who share power with the LDP, force their voters to vote along party lines and use cult-like mind control tactics among their church members in the "Sokka Gakkai" to elicit uncritical compliance. The LDP political machinery that gets supporters out to vote for LDP candidates on any election day is also legendary in Japan. It is unfortunate that more young voters did not turn out as they may have been swayed by Utsunomiya's arguments, but for whatever reason, either due to apathy or the heavy snow, they stayed home ("Voters in their 20s," 2014).

The Role of NHK and Other Media Outlets

The role of NHK (Nippon Hoso Kyokai – Japan's official broadcasting corporation) and other broadcasting stations' contribution to the nuclear village's subversion of the Tokyo election is also important to note. In a clear case of governmental and corporate censorship, at least two commentators were censored from expressing their views just over a week before the election. Professor Toru Nakakita is a veteran radio show commentator and professor of economics at Toyo University. He was formerly with the Foreign Ministry and served as the deputy chairman of the Council for the Asian Gateway Initiative in the first Shinzo Abe Cabinet. Nakakita was told by the NHK director of a show to change the subject of his commentary, which was critical of costs consumers would bear for the resumption of nuclear reactor operations in Japan. In other words, he was not offering up the usual pro-nuclear blather that the nuclear village preferred the mainstream media to embrace. Nakakita was told by NHK (increasingly under the political pressure and influence of the Abe administration) to postpone his commentary until after the election because it "would affect voting behavior." Nakakita quit his job in protest.

Prof. Nakakita was quoted as saying,

The director kept insisting that people vote based on 'impressions.' But I wonder if it's OK to say we can talk about contentious issues at length only after the election. What if I had talked about welfare? Wouldn't that have affected the voting behavior? The media should choose various issues especially during the campaign [...] If they don't, voters will go to the polls with no information to base their judgments on. Isn't it the mission of the news organizations to have the guts to give more information to the public? ("Censorship, crackdown," 2014)

Another well-known broadcaster, Peter Barakan, is a UK national who has been in Japan for about 40 years and therefore broadcasts in Japanese. Barakan mentioned on a private FM radio station that he had been pressured by two other broadcasting stations to steer clear of nuclear power issues on his programs until after the Tokyo gubernatorial election. The two broadcasting stations were not named, but it is possible that one of them was NHK given that he regularly appears on both NHK radio and TV.

This is further proof that the tentacles of the nuclear village reach deeply into all segments of the media. The fact that Prof. Nakakita was told to stay clear of nuclear issues by his director suggests that the word had come down from the top levels of NHK to radio and TV directors to keep nuclear issues off the air until after February 9 (Otake, 2014).

Interestingly, the NHK Chairman, Katsuto Momii, was summoned to the Diet budget committee on January 31 to answer questions about recent statements, perceived by some Diet members to be biased or controversial, he had made at his inaugural press conference on January 25. The proceedings of this committee are broadcast live on NHK TV. One statement concerned wartime "comfort women," about which Mr. Momii had stated that "[a]ll countries did it," and in another statement Mr. Momii had said with regard to broadcasting policy on the international service of NHK, "We cannot very well say 'left' when the government says 'right,'" which may have sounded to some Diet members as if Mr. Momii was intending that NHK would be a faithful echo of government positions. Article 4 of the NHK Broadcasting Law that provides for politically fair and unbiased broadcasting was invoked during the questioning of Mr. Momii during the budget committee session ("NHK Chairman," 2014).

However, the cases of Prof. Nakakita and Peter Barakan were not mentioned. Surely this is not because no one in the room knew about them; both men are extremely well known in Japan and the gag stories had been in the news in the past few days. So it seems "comfort women" and government-leaning broadcasting can be talked about in parliamentary committees, but when it comes to gagging possible comments on nuclear issues on Japan's radio and TV before an election, a censorship issue Diet members ought to be very sensitive about, they cannot find the courage to address it. Why not? Well, the Diet members can't very well mention nuclear issues on live TV only about a week before the Tokyo gubernatorial election, can they? Otherwise, that would betray the role elucidated by the JCP earlier in this chapter that LDP politicians are just pre-programmed sock puppets for the nuclear industry. Other than the notable exception of Taro Kono, if they don't toe the party and big business line they will lose campaign funding and support.

Indeed, censorship of the nuclear accident is well documented in the period prior to the recent election. One of the authors, Wilcox, attests to personally being told not to write for a college journal on the nuclear topic as it was deemed "too sensitive" (Wilcox, 2011). In 2012, at a meeting of Japanese journalists many complained that they would be fired if they wrote about the dangers of radioactive contamination in Japan ("Issues of," 2013).

How many voters did the nuclear village reckon would switch to Mr. Utsunomiya (or Mr. Hosokawa, the decoy duck) if a few commentators were to mention nuclear power on the radio or TV in the days before the election? It's not as if they were going to say, "Vote for Mr. U!" after all. As in the quote from Prof. Nakakita above, open discussion of all the issues is the stated mission of the media, without which the voting public has little on which to base voting decisions – so of course voting behavior is affected; that's what the media are for since democracy depends upon the free flow of information, without which it cannot be called democracy. Thus the nuclear village attempts to plug its finger in every little anti-nuclear leak in the dike that appears in the media (whether or not immediately prior to elections), while using its influence and large media budgets to ensure that the pro-nuclear power message reaches the public through advertising and program sponsorships (Meyer, 2011).

In all fairness, the winner of the election, Mr. Masuzoe, did cleverly put the nuclear issue on the back burner and focused on more immediate concerns to voters ("Coalition shuns," 2014). This is an understandable and winning strategy. It also underlines the philosophical problem of environmental issues which are large and long term, biogeographical and intergenerational in their implications, and generally beyond the understanding of the average person, who is deliberately taught to not think critically but unquestioningly accept guidance from "the experts" on TV. Or perhaps it is just that large numbers of Japanese voters simply go out and vote for the candidate the media "suggest" will win just for the satisfaction of having picked the winner, who happens to have an "open" face and a pleasant smile, like Mr. Masuzoe.

The lack of journalistic honesty and coherence from the media leads to muddled thinking on the part of uneducated or naive voters. Witness the schizophrenic, nihilistic, yet brutally honest comments from this young man:

> Do you think the nuclear issue should be the focus of the Tokyo election?
> Student, 20 (Japanese), "I don't think nuclear power should play a role, because in reality the issues will never be solved."

At the other end of the spectrum is a stubborn old man who gives no reason for his answer:

> Electrical engineer, 65 (Japanese), "I don't think the nuclear issue should feature as part of the debate surrounding the upcoming election." (Buckton, 2014)

Since the authors live in Tokyo or to the north of the capital, we do feel that nuclear issues affect us directly. We would have supported any of the anti-nuclear candidates if they had had a chance to win. In theory, Hosokawa would have had the best chance (had he been genuine in his run), given that the poker game is rigged by the LDP "old boys' club" of which he was once a mildly progressive member.

Recall the case of Ron Paul, now retired representative from the state of Texas, in his strongly articulated reformist role as a Republican Party presidential candidate in the 2012 US election. Paul called for the closing of the global US military base empire and had common sense policies to rebuild America, but was roundly laughed at, ridiculed and ignored by the New World Order corporately-controlled media for his principled and reasonable positions.

A Man of Integrity

Japan's nuclear cartel, those vested interests in the nuclear industry and the heavy industries that rely on the electricity, along with the corrupt politicians and criminal gangs (*yakuza*) that provide the muscle to act as enforcers, would never allow a progressive lawyer like Utsunomiya into such a prominent office. If he did manage to win, they could find a million ways to derail his program, ranging from using the media to carry out negative coverage to the creation of fake scandals or resorting to bureaucratic intransigence in order to block his reforms, not to mention threats of violence given the construction/nuclear industry and LDP ties to organized crime (McNeil, 2009; Wilcox, 2012).

Utsunomiya ran for Tokyo governor in 2012 on an anti-nuclear platform but lost, and then reentered the race again for the 2014 election. He ran with the support of the JCP on both occasions. Due to his concern for their economic struggles, an important part of his power base in Tokyo is young mothers. This was evident by the number of baby stroller mothers who turned out for Utsunomiya's speeches during the last election. The right wing reactionary and former health minister, Masuzoe, on the other hand, who has never spoken out against radiation dangers and is just a shill for big business, has cut old age pensions in favor of giving subsidies to TEPCO.

> The former president of the Japan Federation of Bar Associations, Kenji Utsunomiya.... His main political platform is to abolish nuclear power in the country, beginning with Tokyo.... Utsunomiya said that he is willing to work with other municipalities so that they can pressure the central government

to finally and fully abolish nuclear power. He is also working to provide support to the victims of the nuclear disaster in Fukushima, like providing housing for them in Tokyo. He also said that since Tokyo is the biggest consumer of electricity in Japan and is the biggest shareholder in Tokyo Electric Power Co, it is their responsibility to help and support the victims of the nuclear disaster. Utsunomiya.... has made it his life's mission to fight poverty and has also vowed to expand employment opportunities for the lower classes and improve their welfare and medical care. (Torres, 2012)

However, even if Utsunomiya had won, there could have been problems. He's a lawyer, not a politician in the conventional mold, and very clean, with his heart in the right place. Could he have persuaded the Tokyo assembly to work with him? Our feeling is that the assembly would have been obstructionist – as they very effectively were when it became necessary to oust Inose.

Now we have Masuzoe and the assembly will cooperate, but what will he do? Most likely he will make evasive statements about nuclear reactor restarts (while cooperating behind the scenes with the Abe drive for as much nuclear power as soon as possible) and otherwise pander to sundry business interests. Either way, ordinary people LOSE.

It should also be remembered that Hosokawa and Koizumi (H/K) share political origins with Abe and there is unlikely to be any real anger or contradiction between them, which is why the election turned out very conveniently for the LDP. Since H/K do not effectively have significant future political lives, they have become much-appreciated martyrs to the cause and will be allowed to drift into comfortable retirement for the service they have rendered for Abe and his crew in the election and anything they may yet do along the same lines to confuse voters about the nuclear issue.

By contrast to Utsunomiya, there is evidence that Hosokawa and Koizumi are not as committed to true reform. Where have H/K been for the last three years with their anti-nuclear platform? Were they simply somewhat behind the learning curve or just out to play the spoiler on behalf of Abe? Hosokawa's rhetoric about renewables and energy conservation may have sounded well and good, but Hosokawa waffled and contradicted himself on the issues to make himself more appealing to pro-nuke voters, originally saying, correctly in our opinion, that Japan should not have attempted to host the Olympics, but later taming his rhetoric with vapid slogans about creating a "new Tokyo and Japan" by 2020.

Hosokawa was, as noted above, a late comer to the campaign who had not made his pledges public, but had found support from some anti-nuclear activists who "urged Utsunomiya, an opponent of nuclear power, not to run and thus avoid a split in the no-nukes vote." Utsunomiya, however, would not drop out without more clearly stated goals from Hosokawa, and noted that it was an "abnormal election" because the candidates had already decided to run but were not able to have a policy debate (Kameda & Yoshida, 2014).

The Winter of Mal-Dissent

The anti-nuke movement offered hope to Japan during the summer of 2012 (Hayashi et. al., 2012), but a year later had exhausted itself (Samuels, 2013). The Tokyo election is an example of failed democracy in the context of the nuclear power paradigm. The Masuzoe victory was due to a fractured opposition vote and LDP power over the media ("Masuzoe scores," 2014; "NHK governors," 2014).

Why did the street movement and ballot box method fail? Cooptation and infiltration are tried and true tactics of governments to defang dissident groups by fragmenting and discrediting their agendas. The most famous example in the US was "Co-intel-pro" (i.e., the FBI's counterintelligence program) ("Cointelpro," 2014).

In Japan the police and the army spy on protestors and keep records just in case they need to intimidate them. But usually they simply buy off protestors by offering them trinkets and grant money to steer their campaigns in less challenging directions. In the West, this is the George Soros grand strategy of mass manipulation, liberal totalitarianism and fake democracy (Vltchek, 2014; Sigursdatter, 2014).

NGOs and nonprofits know the limits of what they are allowed to do. Furthermore, the anti-democratic "State Secrets" law that passed in 2013 has further clamped down on the activities of anti-nuclear activists and due-diligence writing journalists (Mie, 2013).

What does that leave for people who are still worried about nuclear power restarts under the Abe government? How can they show their passionate opposition? Perhaps they could take a page out of the recent anti-nuclear victory in Taiwan where strong public opposition temporarily stopped construction of a nuclear power plant, pending a referendum, on that small, seismically active island ("Construction halted," 2014). In Japan's case, due to the power of the nuclear lobby, the only other answer for concerned Japanese people may be to carry out stealth, non-violent pro-active protest by "going off the grid" and starting local, micro-electricity grids that rely on solar power (See chapter by Boys and Wilcox in this volume; Maharrey, 2014).

Japan Should Adopt a Renewable Energy Future

The nuclear issue is a political and not a technical problem. Notwithstanding the fact that we don't know what Japanese society and the economy will look like in 2050, there are plenty of data and studies to suggest that Japan *could* meet 100 percent of its electricity demands by 2050 (Johnston, et. al., 2013). The key issue, however, is how to create proper incentives for investment ("12 Insights," 2013). As one foreign professor who is teaching and residing in Japan points out, a broad subset of German social groups chose to abandon nuclear power and it is a decision that would surely be shared by a majority of Japanese, were they allowed to make it:

> One country that many people have looked to is Germany and its decision to embark on an 'energy revolution' (Energiewende). For anti-nuclear supporters, Germany is an example of what Japan should be doing. But the lesson that should be taken from Germany is not the decision itself, but the process by which it was reached... [A]n Ethics Commission on Safe Energy Supply, composed of a cross section of German society with representatives from politics, industry, academia and religion [which] collectively reflected on what was best for the country and its future. (Hobson, 2014)

This Ethics Process must be made in defiance of the deeply embedded oligarchy that has, in reality, sternly ruled the country for centuries. The oligarchy is not easily forgiving of dissenting opinions nor open to revelations about the future that do not reverberate with its image of "how things should be" as they conveniently reap the benefits and call the shots. Japan faces an enormous political challenge toward denuclearization, and this is also why spontaneous, sustainable and massive civil disobedience in favor of denuclearization is not likely to occur in the near term. Only through the hard work of grassroots activism and education in conjunction with the natural atrophying of the nuclear project, as the monetary well runs dry and the realization among the politicos that the nuclear dream of power that is "too cheap to meter" dies a tortuous and ugly death, will things eventually change for the better.

"How beautiful life is and how sad! How fleeting, with no past and no future, only a limitless now" (Clavell, 1980).

References

Blatant behind the scenes stratagems in this crucial period. (2014, February 1). *Tokyo Shimbun*, pp. 28-29.

Boys, A.F.F. (2000). Food and Energy in Japan: How will Japan feed itself in the 21st century? PDF Retrieved from http://www9.ocn.ne.jp/~aslan/21fee.pdf

Buckton, M. (2014, February 5). Do you think the nuclear issue should be the focus of the Tokyo election? T*he Japan Times*. Retrieved from http://www.japantimes.co.jp/community/2014/02/05/voices/do-you-think-the-nuclear-issue-should-be-the-focus-of-the-tokyo-election/#.Uvp8ZqWyOfl

Busby, C. (2006). *Wolves of Water*, Green Audit Books.

Censorship, crackdown on Fukushima information. (2014, January 31). *Nuclear-news*. Retrieved from http://nuclear-news.net/2014/02/02/censorship-crackdown-on-fukushima-information-professor-resigns-from-broadcasting-in-protest/

Clavell, J. (1980). *Shogun: A Novel of Japan*. Dell Books. Retrieved from http://www.amazon.com/Shogun-Novel-Japan-James-Clavell/dp/B000HFWEBG

Coalition shuns nuclear issue in Tokyo election; Masuzoe still in front. (2014, February 3). *The Asahi Shimbun*. Retrieved from http://ajw.asahi.com/article/behind_news/AJ201402030074

Cointelpro: The Sabotage of Legitimate Dissent (2014). *What Really Happened.* Retrieved from ttp://whatreallyhappened.com/RANCHO/POLITICS/COINTELPRO/cointelpro.php

Cohen, M. & McKillop, A. (2012). The Doomsday Machine: The High Price of Nuclear Energy, The World's Most Dangerous Fuel. Palgrave.

Construction halted at Taiwan nuclear plant after protests. (2014, April 27). Nuclear Power Daily. Retrieved from http://www.nuclearpowerdaily.com/reports/Construction_halted_at_Taiwan_nuclear_plant_after_protests_999.html

Duke, D. (2014, January 2). Financial Swindling "Natural Outgrowth of Jewish Community"—"Jewish Journal." *DavidDuke.com*. Retrieved from http://davidduke.com/jewish-extremist-financial-swindling-natural-outgrowth-jewish-community-jewish-journal-editor-confesses/

Electric companies urge LDP members to promote nuclear power (2014, January 29).

Japan Press Weekly. Retrieved from http://www.japan-press.co.jp/modules/news/index.php?id=6882

Favole, J. A. & Tennille, T. (2011, March 15). Obama Stands by Nuclear Power. *The Wall Street Journal*. Retrieved from http://online.wsj.com/news/articles/SB10001424052748703363904576200973216100488

Gene 'Chip' Tatum: What really happened to MH370? (2014, April 14). *Kevin Barrett's Truth Jihad Radio, No Lies Radio*. Retrieved from http://noliesradio.org/archives/80940

Gundersen, A. (2014, February 3). Fukushima's Refugees: Why Have They Been Abandoned? [Video Podcast]. *Fairewinds Energy Education*. Retrieved from http://fairewinds.org/media/fairewinds-videos/fukushimas-refugees-abandoned

Gundersen, A. 2014, February 13. New TEPCO Report Shows Damage to Unit 3 Fuel Pool MUCH Worse Than That at Unit 4. [Video Podcast]. *Fairewinds Energy Education*. Retrieved from http://fairewinds.org/media/fairewinds-videos/new-tepco-report-shows-damage-unit-3-fuel-pool-much-worse-unit-4

Hayashi, Y. & Sekiguchi T. (2012, July 27). Political Clout of Japan's Anti-Nuke Movement Tested. *The Wall Street Journal / Asia.* Retrieved from http://online.wsj.com/news/articles/SB10000872396390443343704577552102 304673574?mg=reno64-wsj&url=http%3A%2F%2Fonline.wsj.com%2Farticle%2FSB1000087239639 044334370457755210230467574.html

Hobson, C. (2014, February 7). Lessons for fixing Fukushima. *The Japan Times.* Retrieved from http://www.japantimes.co.jp/opinion/2014/02/07/commentary/lessons-for-fixing-fukushima/#.UvaK06WyOf0

Hongo, J. (2011, March 27). Signs of disaster were there to see. *The Japan Times.* Retrieved from http://www.japantimes.co.jp/news/2011/03/27/news/signs-of-disaster-were-there-to-see/#.UvP_FaWyOf0

Issues of Radioactive Exposure are Considered Taboo on Japanese Media. (2013, February 13). *Independent Media Symposium.* Retrieved from https://www.youtube.com/watch?v=NHtbi1Q4aZ8

Japan's U-turn on the zero option for nuclear energy. (2012, September 22). *Idaho Samizdat.* Retrieved from http://djysrv.blogspot.jp/2012/09/japans-u-turn-on-zero-option-for.html

Johnston, E. & Institute for Sustainable Energy Policies (Eds.). (2013). Renewables Japan Status Report 2012; A Vision of Japan's Renewable Energy Landscape in 2050. *Fresh Currents.* Retrieved from http://download.freshcurrents.org/FreshCurrents2012-final.pdf

Kameda, M. & Yoshida, R. (2014, January 22). Hosokawa to play Tokyo nuke card. *The Japan Times.* Retrieved from http://www.japantimes.co.jp/news/2014/01/22/national/hosokawa-to-play-tokyo-nuke-card/#.Uvp1HKWyOf1

Kameda, M. (2014, February 6). Activists fear gubernatorial-race standoff could split anti-nuclear vote. *The Japan Times.* Retrieved from http://www.japantimes.co.jp/news/2014/02/06/national/activists-fear-gubernatorial-race-standoff-could-split-anti-nuclear-vote/#.UvpcoqWyOf1

Kingston, J. (2012, September 10). Japan's Nuclear Village. *The Asia-Pacific Journal, Vol. 10, Issue 37, No. 1.* Retrieved from http://www.japanfocus.org/-Jeff-Kingston/3822#sthash.1Pzbb4C2.dpu http://www.japanfocus.org/-Jeff-Kingston/3822

Maharrey, M. (2014, April 18). The Tougher Path to Freedom: Non-Violence. *Activist Post.* Retrieved from http://www.activistpost.com/2014/04/the-tougher-path-to-freedom-non-violence.html

Masuzoe scores victory in Tokyo gubernatorial race. (2014, February 9). *The Japan Times.* Retrieved from http://www.japantimes.co.jp/news/2014/02/09/national/masuzoe-set-for-victory-over-anti-nuclear-foe-in-tokyo-gubernatorial-poll/#.UvlPokJdVbw

Masuzoe Yoichi. *Wikipedia.* Retrieved from
http://ja.wikipedia.org/wiki/%E8%88%9B%E6%B7%BB%E8%A6%81%E4%B8%80

McNeill, D. (2009, June 9). *The Japan Times.* Rumpus on campus: Prestigious
university in Tokyo has become a battleground in a war over freedom of
political expression. Retrieved from
http://www.japantimes.co.jp/community/2009/06/09/issues/rumpus-on-campus/#.U1cejqXjeuc

Meyer, C. (2011, May 27). Japan's Nuclear Cartel: Atomic Industry Too Close to
Government for Comfort. *Speigel Online International.* Retrieved from
http://www.spiegel.de/international/world/japan-s-nuclear-cartel-atomic-industry-too-close-to-government-for-comfort-a-764907.html

Mie, A. (2013, December 6). Diet enacts controversial state secrets bill. *The Japan
Times.* Retrieved from
http://www.japantimes.co.jp/news/2013/12/06/national/secrets-bill-poised-for-passage/#.UvliSaWyOf1

Mogi, C. & Ujikane K. (2014, January 26). Japan Has Record Trade Deficit as
Rising Fuel Bill Hurts Growth. *Bloomberg News.* Retrieved from
http://www.bloomberg.com/news/2014-01-26/japan-record-annual-trade-deficit-shows-import-drag-on-recovery.html

Nadesan, M. H. (2013). *Fukushima And The Privatization Of Risk.* Palgrave/Pivot.

NHK Chairman makes statement, apology. (2014, February 1). *Tokyo Shimbun,* p.3.

NHK governors back Abe agenda, minutes reveal. (2014, February 9). *The Japan
Times.* Retrieved from
http://www.japantimes.co.jp/news/2014/02/09/national/nhk-governors-back-abe-agenda-minutes-reveal/#.UvlRQ6WyOf0

Nuclear Village. *Wikipedia.* Retrieved from
http://ja.wikipedia.org/wiki/%E5%8E%9F%E5%AD%90%E5%8A%9B%E6%9D%91

Nuke Info Tokyo (archive 1987 - 2014). *Citizens Nuclear Information Center.*
Retrieved from http://www.cnic.jp/english/newsletter/index.html

Obayashi, Y. (2014 April 20). Fukushima No. 1 boss admits water woes out of
control; Abe told an Olympic Committee meet situation was under control.
The Japan Times. Retrieved from
http://www.japantimes.co.jp/news/2014/04/20/national/fukushima-no-1-boss-admits-water-woes-out-of-control/#.U1bDta1dWZJ

Otake, T. (2014, January 22). Barakan says broadcasters told him to avoid nuclear
issues till after poll. *The Japan Times.* Retrieved from
http://www.japantimes.co.jp/news/2014/01/22/national/barakan-says-broadcasters-told-him-to-avoid-nuclear-issues-till-after-poll/#.UwYkq6WyOf1

Samuels, L. (2013, March 18). Japan's Anti-Nuclear Activists Losing Ground Since Fukushima Disaster. *The Daily Beast*. Retrieved from http://www.thedailybeast.com/articles/2013/03/18/japan-s-anti-nuclear-activists-losing-ground-since-fukushima-disaster.html

Sigursdatter, I. (2014, April 14). Norway: Happiest Country on Earth Myth. *Red Ice Creations*. Retrieved from http://www.redicecreations.com/radio/2014/04/RIR-140414.php

Solzhenitsyn, A. (1973). *The Gulag Archipelago: 1918 - 1956*. Westview Press.

Stuxnet worm hits Iran nuclear plant staff computers. (2010, September 26). *BBC*. Retrieved from http://www.bbc.co.uk/news/world-middle-east-11414483

Tabuchi, H. (2011, July 13). Japan Premier Wants Shift Away From Nuclear Power. *The New York Times*. Retrieved from http://www.nytimes.com/2011/07/14/world/asia/14japan.html?_r=0

Tokyo assembly votes down nuclear referendum ordinance. (2012, June 21). *The Asahi Shimbun*. Retrieved from http://ajw.asahi.com/article/behind_news/politics/AJ201206210045

Tokyo governor contradicts PM's claim that Fukushima is "under control". (2013, September 21). *Japan Today*. Retrieved from http://www.japantoday.com/category/national/view/tokyo-governor-contradicts-pms-claim-that-fukushima-is-under-control

Tokyo gubernatorial election, 2014. *Wikipedia*. Retrieved from http://en.wikipedia.org/wiki/Tokyo_gubernatorial_election,_2014 and http://ja.wikipedia.org/wiki/2014%E5%B9%B4%E6%9D%B1%E4%BA%AC%E9%83%BD%E7%9F%A5%E4%BA%8B%E9%81%B8%E6%8C%99

Torres, I. (2012, November 13). Former lawyer to run for Tokyo governor, vows to get rid of nuclear power. *Japan Daily Press*. Retrieved from http://japandailypress.com/former-lawyer-to-run-for-tokyo-governor-vows-to-get-rid-of-nuclear-power-1318274/

Trento, J. (2012, April 9). United States Circumvented Laws To Help Japan Accumulate Tons of Plutonium. *DC Bureau*. Retrieved from http://www.dcbureau.org/201204097128/national-security-news-service/united-states-circumvented-laws-to-help-japan-accumulate-tons-of-plutonium.html

12 Insights on Germany's Energiewende. (2013, February). *Agora Energiewende*. Retrieved from http://www.agora-energiewende.org/topics/the-energiewende/detail-view/article/12-insights-on-the-energiewende/

Utsunomiya 1st to declare his candidacy for Tokyo governor. (2013 December 29). *Japan Today*. Retrieved from http://www.japantoday.com/category/politics/view/utsunomiya-1st-to-declare-his-candidacy-for-tokyo-governor

Utsunomiya vote total. (2014, February 11). *Tokyo Shimbun*, p. 1.

Vltchek, A. (2014, February 3). How the West Manufactures "Opposition Movements." *Counterpunch*. Retrieved from http://www.counterpunch.org/2014/02/03/west-manufactures-opposition-movements/

Voters in their 20s show little interest in Tokyo gubernatorial race: poll. (2014, February 3). *Mainichi*. Retrieved from http://mainichi.jp/english/english/newsselect/news/20140203p2a00m0na009000c.html

Walter Burien explains how non-disclosure of CAFR data is criminal business. (2012, July 3). *Washington's Blog*. Retrieved from http://www.washingtonsblog.com/2012/07/walter-burien-explains-how-non-disclosure-of-cafr-data-is-criminal-business.html

Wasserman, H. (2014, February 2). 50 Reasons We Should Fear the Worst from Fukushima. *EcoWatch*. Retrieved from http://ecowatch.com/2014/02/02/50-reasons-fear-fukushima/

Wilcox, R. (2011, June 29). Censorship in Japan. *Dissident Voice*. Retrieved from http://dissidentvoice.org/2011/06/censorship-in-japan-the-fukushima-cover-up/

Wilcox, R. (2012, August 28). The Nuclear Mafia Derails Democracy In Japan. *Rense.com*. Retrieved from http://rense.com/general95/nuclearmafia.html

Wilcox, R. (2013, April 14). Conference Highlights Consequences of Fukushima and Nuclear Absurdity. *DiaNuke.org*. Retrieved from http://www.dianuke.org/conference-highlights-consequences-fukushima-and-nuclear-absurdity/

Wilcox, R. (2014, April 12). Japan's Radioactive Potemkin Village. *Rense.com*. Retrieved from http://www.rense.com/general96/jpsradioctv.html

Zimmerman, P. (2009). *A Primer in the Art of Deception*, Paul Zimmerman.

DISPOSSESSION

Fukushima: The Dispossession of Reality by Physics

Christopher Busby

Hegel remarks somewhere that all facts and personages of great importance in world history occur, as it were, twice. He forgot to add: the first time as tragedy, the second as farce.

Marx *The 18th Brumaire of Louis Bonaparte* 1852

Introduction

Although official scientific experts reassure and deny, we know what will happen to the people of Japan after the Fukushima disaster by analogy, by induction. We have seen and measured what has happened to the people of those ex-Soviet and other territories in Europe (and further afield) exposed to the same radioactive fallout. The parallels between the events are frighteningly comparable and were so from the very beginning. To anyone watching the Fukushima explosions it was starkly evident that what was happening involved massive releases of the radioactive contents of the four reactors and their associated spent fuel pools. This was a serious event. The violence of the explosions and the nature of un-cooled nuclear reactors made it obvious that the pressure vessels were breached and molten fuel was out in the atmosphere, out in the ground. This has since been confirmed from a whole range of clues, which I will return to. But from the beginning, and despite the fact that intercepted phone calls and leaked documents later were to show that the authorities knew quite well what was happening, the public were bombarded with disinformation by the media. Scientists and experts appeared with facile and dishonest placatory messages. Nothing like Chernobyl, we heard. Indeed, nothing like Three Mile Island: the events barely registered on the international nuclear accident scale (the what?). The radioactivity was blown safely out to sea. Doses to the public were barely worth mentioning. And so forth. I think I was the first scientist saying how bad it was, when I appeared on both the BBC and ITV on 12th March 2011. I was soon shifted out of the picture in favour of the stooges, but I slipped in before the media gate crashed shut. As luck would have it I was just passing through London on my way from France to Wales when they needed an expert in the studio, and the interviewer was one that knew me from way back. Serendipity. And it became clear after a few weeks that everything I had said was correct. But the media reportage rapidly developed in the same way as it did (and has continued to do) after Chernobyl. So the truth war battle lines also are drawn up: the governments, the nuclear industry, the media and tame nuclear industry scientists on the one side, the few independent scientists and experts on the other. And in the middle, the sick and dying, abandoned by their governments in Japan, as in Belarus.

In this contribution I will leave the minute description of the causes and the sequences of the disaster to others and will concentrate on two things. First, the radiation exposures and the effects that they will have on health in the local exposed population. These will be catastrophic. I will also look at the effects on those living in more distant areas contaminated by the releases, indeed as far away as the USA, more than 4000 miles downwind. Second, I will outline the reasons for the massive cover-up and the disinformation operation. This process represents the dispossession of reality of the title, which I will argue is the offspring of a fatal (for the rest of us) marriage between the particular methodologies of reductionist physics with the needs and desires of the nuclear and military complex. It is impossible for the ordinary person to understand the diametrically opposed views there are of the health outcomes of Fukushima without understanding the significance of the event for the official view of radioactive pollution. Therefore an introduction to the effects of radiation and how they have been historically measured and assessed is necessary before one can make sense of Fukushima.

From Exposure to Ill Health: The Risk Model

My tool for predicting and explaining the health effects is largely the radiation risk model of the European Committee on Radiation Risk (ECRR), which I helped develop, and will introduce and explain [ECRR 2010]. But before I begin I will introduce my second point by discussing the way in which, in the field of "radiation protection", the public and their political leaders have been led astray in the last 60 years by a system of dishonest scientists and their dishonest organizations, which remains in place today and which retains a death grip on the perception of risk from low doses of ingested and inhaled radioactivity. This bogus system was constructed during the Cold War in order to support the testing of nuclear weapons. It constrained the doctors of the World Health Organization (WHO) in 1959 through an agreement signed with the International Atomic Energy Agency (IAEA) to leave research on health to the atom scientists. It remains in place through the influence and power of the international nuclear and military complex. Its risk model underpins all laws relating to exposure limits. But it is a very simple matter to show that it is wrong, and as a result tens of millions have died and tens of millions will die in the future. And indeed the evidence is there to see, published in the scientific peer review literature, but still nothing is done about what is arguably the greatest public health scandal in recorded history. How can this be? But to begin, what does "radiation risk model" mean? Why does it matter?

From the discovery of X-rays and radioactivity, the last century saw increasing realization that radiation was genotoxic. Radiation exposure created genetic damage, which manifested itself as cancer in those exposed and their progeny, congenital damage (miscarriage, stillbirth, birth defects) and by the 1960s clear evidence of a general increase in all diseases with associated loss of lifespan: a kind of non-specific ageing. Experiments with X-rays and fruit flies showed clearly that chromosomes were altered by radiation, and so the long term effects on genetic integrity, mutations and effects in offspring became an issue. It became clear that workers and the public should have legal protection against being exposed to radiation. If there were laws regulating exposure to chemical poisons like cyanide, arsenic, cadmium and lead, then there must be developed similar rules for radiation.

The problem was (and is) that whilst you can measure chemical exposures and limit the amount of lead or strychnine in grams that can be allowed to be ingested or inhaled, radiation is not amenable to such an approach. It is invisible and is measured only with complex machinery. Up to the 1940s, the way of measuring radiation exposure was using the humans themselves as detectors, actually quite the most accurate method for immediate exposure, though not much use for cancer that develops after twenty years. An exposure to a chemical poison can be described by the term "dose", which is a defined amount of ingested material in grams, like a dose of aspirin. The term "dose" came to be used in early radiation protection to describe the quantity of radiation which would cause the health effect being examined. So in the 1920s initial burning of the skin by an X-ray was termed the "minimal erythemal dose". But by the 1940s, this approach was abandoned and physicists (who like to simplify things) invented the idea of "absorbed dose", which was defined as absorbed energy (Joules, ergs) per unit mass (kilograms, grams). This is very important, so please try to get this straight right here. It is the trick that is the key to understanding all that follows. Radiation, X-rays, gamma rays, etc. travel though you, but some of the energy is absorbed (otherwise X-rays would not provide pictures of your insides). Dense material like bone absorbs more, so the doctor can see the fracture in the bone on the X-ray film. The amount of energy (in Joules) absorbed by a <u>kilogram</u> of living tissue is the "absorbed dose". This was laid down as law in 1947 and adopted by a new organization, called the International Committee for Radiological Protection (ICRP) in 1952. To allow for the increased dangers from alpha radiation due to its more highly ionizing short range tracks, a multiplier of 20 was used to create a variation of this absorbed dose called the "equivalent dose", but for the purposes of my argument this account is good enough.

So the "risk model" is a mathematical relation between the amount of radiation that a person receives, defined by their absorbed dose, and the amount of ill health that that dose produces defined as cancer. It is entirely analogous to the risk model that predicts the dose of lead or sodium cyanide necessary to make you ill or to kill you, or more appropriately for some chemical carcinogens (benzpyrene, aflatoxin), which will cause genetic damage and cancer long after they were ingested.

Whilst the physicists approach to measuring radiation in the body was fine for X-rays and for gamma rays (which are a sort of high energy, invisible form of light), the method was not really applicable to a different kind of radiation exposure. This was internal exposure to the new radionuclides that fissioning of the Uranium atom had created after 1945. For the first time in evolution,

76

radioactive forms of natural elements appeared in the biosphere. These included all the radionuclides that appeared after Fukushima and Chernobyl and are now quite well known to everyone e.g. Caesium-137, Strontium-90, Plutonium-239, Iodine-131. The Russian writer Alla Yaroshinskaya included a new isotope Lie-86 in the list of Chernobyl releases, and in the same spirit we could perhaps include the radionuclide Coverup-11 from Fukushima. However, in the period 1950-1963, these substances were created in all the many atmospheric nuclear tests carried out by the US, the Soviets and Great Britain, and rained down all over the world. Although in absorbed dose terms the levels were small, about 1 milliSievert (mSv) there was a clear increase in infant mortality in the USA and UK (reported in the peer review literature and twenty years later the cancer epidemic (which has touched all of you) began. Strontium-90 from weapons fallout is now in the bones of everyone on earth from Baffin to Brisbane, from Kut to Kandahar.

From the start of the ICRP, there was a big argument about how to incorporate exposure to these "internal radionuclides" into radiation protection: whether they could be described through the concept of "absorbed dose". Committee Two, headed by Karl Z Morgan, argued they could not. We now know he was correct. However, the requirements of the military in the cold war period prevailed; Morgan resigned and the methodology of absorbed dose was transferred seamlessly across to internal radionuclide exposures. The energy released in each radionuclide decay, wherever it was on the molecular scale, was diluted into the kilogram of material in which the nuclide was embedded to give energy density in Joules per kilogram. One Joule per Kilogram is one Gray (or one Sievert). One milliSievert is one thousandth of a Joule per kilogram. This was the mistake that has echoed down to Fukushima and is the reason for all the disagreements and misdirections which are here with us today, killing the children in Japan, those exposed to Chernobyl releases, populations of Iraq and the Balkans exposed to Uranium weapons, those living downwind of contaminated nuclear sites, and all those (which is everyone) exposed to the weapons tests fallout from the 1960s.

Reductionist Physics dispossessed the human race of its ability to independently assess reality, and not for the first time (relativity, quantum theory, black holes, dark matter, Eddington's table, Higgs Boson). Physicists are dangerous. They should have rotating blue lights fitted. They wield a kind of Emperor's New Clothes authority out of all proportion to their ability to connect with the empirical observations and can deny (and do deny) what is placed in front of them on the basis of mathematical equations. It as if you see a poisonous snake in the park (the bite causes leukemia) and refuse to allow your child to play there. The government scientific advisor says that mathematical equations show that what you thought was a snake was a rabbit and the government forces you to use the park on this basis.

Most of the scientific advisors to government are such people, theoretical physicists. Arguing with them is like trying to beat a cat. They will never address the point but will always refer to some other physicist or theory. The Nobel Prize Chemist and Economist Michael Polanyi realized the danger as long ago as 1958. He compared them to the Azande Witch Doctors thus:

> [For] the stability of the naturalistic system we currently accept, instead, rests on the same logical structure as Azande witchcraft beliefs. Any contradiction between a particular scientific notion and the facts of experience will be explained by other scientific notions. There is a ready reserve of possible scientific hypotheses available to explain any conceivable event. Secured by its circularity and defended by its epicyclical reserves science may deny or at least cast aside as of no scientific interest, whole ranges of experience which to the unscientific mind appear both massive and vital.

> M. Polanyi FRS, *Personal Knowledge,* 1958

But their creation, the Risk Model of the ICRP, a Black Box, can be unpacked and its contents simply examined by Hans Andersen's little boy. And such a forensic review shows unequivocally that it is wildly incorrect because its fundamental basis is wrong. How is it wrong?

For those who want a more scientific account, I refer you to a review I published in 2012 [Busby, 2013]. Absorbed dose is an average energy density in Joules in one kilogram of tissue. But radiation effects are known to occur at the DNA in the cell. Most of the tissue (the non DNA tissue) is unaffected by radiation. All radiation exposures, however they are delivered, create their damaging effect on the genetic makeup of the cell, and do so though the creation of tracks of charged molecular fragments of the breakdown of tissue components, mainly water. These "ions" are reactive and can chemically destroy components of the DNA to cause mutations. Whilst for external radiation, the average concentration of these ions can be agreed to be roughly uniform throughout one kilogram of the tissue, this is not always true for internal radionuclides, which are just like any other chemical substance in the tissue, except that they will suddenly explode, turn into a different element, and create a track of ions. There are a number of ways in which this behavior cannot be incorporated into the "absorbed dose" averaging concept of the ICRP. The main one is for elements like Strontium-90, Plutonium-239 and Uranium-238, 234 and 235, all of which have strong chemical affinity for DNA. In other words, they target the exact material that is the origin of the genetic effects which are the cause of the illnesses. One other way is when the exposure is to micron sized particles, so-called hot particles (but also not-so-hot particles). These were detected after Fukushima in Tokyo, and as far away as Hawaii and California. The ionization density near such particles is very great. Fig. 1 shows a graph of the concentration of Uranium after Fukushima by distance from Japan. This shows the presence of Uranium particles in air, being breathed in. But these particles can be so small they effectively act as a gas and can pass though the skin. An analogy would be comparing sitting in front of a warm coal fire and absorbing X Joules per kilogram over your whole body with eating a red hot coal.

Figure 1 Concentration of Uranium in air by Distance from Fukushima shortly after the explosions. Data from US EPA RADNET.

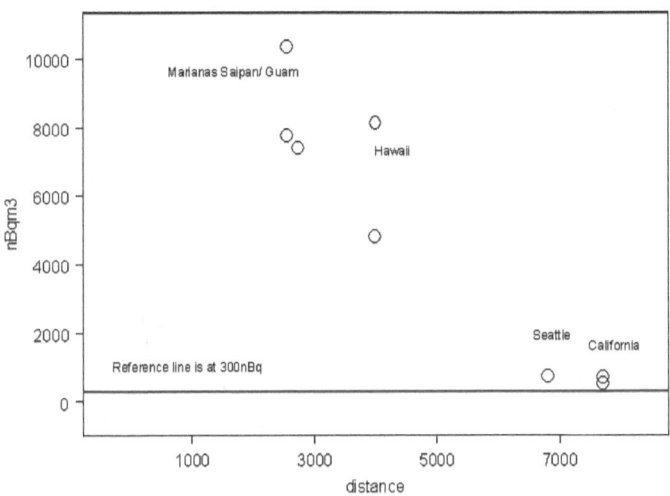

The ICRP risk model is based on comparing the "absorbed doses" of the survivors of the bombing of Hiroshima and Nagasaki with the later incidence of cancer in those exposed compared with cancer in those inhabitants or later entrants who were not exposed to gamma rays at the detonation. In this "life span study" it is assumed there was no internal radiation, but this has now been shown to be a false assumption. Since both the "unexposed" controls and the study group were both exposed to internal fallout radionuclides, especially Uranium from the bomb in the "black rain", the results of the life span study are strictly false and the ICRP risk model wrong. This argument has been made by many, but power and influence have excluded it from policy changes.

What I hope I have succeeded in doing here is to explain how it is that there are two views of the consequences of Fukushima; the official one, based on the ICRP model, and the real one, based on the observations of the effects of exposure to internal radionuclides. From the weapons tests which caused the cancer epidemic of the 1980s though the nuclear site child leukemias, the Chernobyl cover-ups and the Iraq Uranium cancers, from Uranium miners to those living on the edges of radioactively contaminated sites – the Irish Sea, the Baltic sea, Kazakhstan, the Marshall Islands – to the nuclear test veterans and the veterans of the Iraq and Balkans wars, the answer from the authorities is always the same:

the dose is too low. Physicists working for governments are denying what they see with their eyes on the basis of simplistic mathematical descriptions of complex living organisms.

In 1956, when radar was new, the Italian liner *SS Andrea Doria* collided with the liner *SS Stockholm* and sank off Nantucket as a result of depending on radar in fog. This was the first *radar assisted collision*, where an assumption is made that the technology (the model) is telling you the truth without investigating it. In radiation protection we have had such an ongoing radar-assisted collision with reality since 1952, with millions more victims of the process than the *Andrea Doria*. Those who use the ICRP model continue to deny any problems either because of some desperate adherence to their mathematical world-view or in some cases, no doubt, because they are paid to bamfoozle the public. These people have dispossessed us of the evidence of our own eyes. I hope one day to see the physicists who continue to peddle this nonsense sent to jail.

But there is one more relevant point. The ICRP model's success is its simplicity and the assumption that there is indeed such a thing as an absorbed dose which can be calculated and is meaningful for quantifying health risk. This system has been slipped over governments (who, after all, know nothing about science) with ease. It is now adopted and maintained as a given truth handed down by Moses on Sinai by all the official authorities. It is maintained by the IAEA, the WHO (which together constitute the same outfit since 1959), The United Nations Scientific Committee on the Effects of Atomic Radiation (UNSCEAR), the Biological Effects of Ionizing Radiation BEIR committees of the US National Academy of Sciences, and all of the state apparatuses in all western countries that administer the laws and exposure limits which are based on the ICRP model. What is not generally realized is that all these organizations involve largely the same people who rotate around the various agencies and committees. A clue to the machinery behind this operation is the address of the precursor and sister committee to the ICRP, the International Committee for Radiation Units and Measurements, ICRU, which is in Bethesda Maryland, next door to the CIA and the US Environmental Protection Agency. But at the end of the day, the ICRP is "an independent charity", and governments are free to take advice from any other committee of independent scientists. This is why, in 1998, a number of independent scientists, myself included, infuriated by the public health scandal that radiation protection had become after dismissing the real effects of Chernobyl, got together in Brussels and formed the European Committee on Radiation Risk, the ECRR. By 2003, the ECRR had developed a new risk model, one which actually gave the correct answer to the question of the effects of internal exposures, and one which was totally validated in 2004 with the publication of the results of a study in Sweden showing a measurable increase in cancer associated with fallout from Chernobyl. The most recent version of this risk model is available as a free download from the website www.euradcom.org.

By 2009, the ECRR model had been published in English, French, Spanish, Russian and Japanese. The membership of the ECRR had increased to include more than 40 eminent radiation researchers, including many from the ex-Soviet territories who had studied first-hand the effects of internal exposures to fission radionuclides after Chernobyl. They included Prof Alexey Yablokov of the Russian Academy of Sciences, Prof Elena Burlakova, Chief of the Scientific Council on Radiobiology of the Russian Academy of Sciences, Dr Yuri Bandashevsky, famous for his studies of the effects on children's hearts caused by the internal Chernobyl exposures in Gomel, Prof Angelina Nyagu, who studies radiation children's health, Prof Roza Goncharova the geneticist, Prof Inge Schmitz Feuerhake, Prof Mikhail Malko, Prof Shoji Sawada, Prof Michel Fernex, and many other world class eminent scientists. In 2009 at a conference in Lesvos, Greece, 20 of these world class hands-on researchers signed a Declaration (which can be found on the ECRR website) and which concluded:

> (We) urge the responsible authorities, as well as all of those responsible for causing radiation exposures, to rely no longer upon the existing ICRP model in determining radiation protection standards and managing risks.
>
> The Lesvos Declaration (ECRR2010, www.euradcom.org)

I should say that in comparison with ECRR, the various organizations which adhere to the ICRP dose approach are composed mainly of scientists with virtually no research experience whatever. A search of the scientific literature for each name in the list of officials and advisors to the ICRP and the other organizations turns up very little. At the WHO conference in Kiev in 2000, we heard Dr Abel Gonzalez, president of the IAEA and onetime of UNSCEAR saying how there were no measurable effects due to Chernobyl. A search on his research papers in the field reveals nothing at all. His PhD was in some arcane area of physical optics. He has never done any research on radiation and health. On the other hand the ECRR scientists are categorized by an embarrassingly rich list of papers published in the peer review literature.

The ECRR model itself is basically simple. It starts by recognizing that to assess the effects of radioactivity in historic situations, and in order to base exposures on some physical foundation, it has to deal with the concept of "dose". But it then adds weighting factors to the "dose" in the same way as the ICRP model does for alpha particle exposures. The ECRR weighting factors are based on epidemiological studies of internal radioactivity exposures and biophysical considerations of ionization density at the DNA, the target for the health effects. The earliest epidemiological pointer to the magnitude of the weighting factors required to bring the ICRP model in line with the observations was the exposure from weapons fallout and the development of cancer some twenty years later in England and separately in Wales, where the Strontium-90 levels were about twice as great. The multiplier coefficients (in the case of the weapons tests the multiplier is around 300) are also tabulated in ECRR2010 for types of exposure such as exposure to particulates. In addition to these approaches to adjust dose, ECRR also includes assessment of non-cancer effects, non specific ageing and infant mortality. ECRR gives the right answer. What is that answer in Fukushima? Let's see.

Fukushima: The Cancer Yield to 200km

It is ironic and rather creepy that the whole business of cancer risk following exposure, and its assessment, which began in Japan with the Hiroshima bomb and the dose calculations there, is now back in Japan some 70 years later, having officially, at least, remained in exactly the same place. Such and such "dose" of radiation causes such and such increase in cancer. Once we can get a handle on the exposures, we can work up to the cancer yield using the ECRR adjusted doses. What follows is based upon an update of a paper I gave at the joint conference of the German Society for the Radioprotection/ECRR conference held at the Charité Hospital in Berlin in May 2011.

To figure out the cancer yield we require:
1. the doses from each of the radionuclides emitted or some way of assessing these.
2. the population exposed

Since information about the concentrations of the different radionuclides which are contaminating the areas near the plant is not available, a strictly formal application of the ECRR model is not possible. But we can make some assumptions which will give a reasonable idea. There are two approaches to approximating internal dose which both give approximately the same result. We can calculate the area contamination on the basis of the gamma radiation dose rate. Or, we can employ the reports of the International Atomic Energy Agency, IAEA, of contamination level at various distances from the release point. We then assume that the internal ICRP dose is equal to the external reported dose, or that obtained by calculating the dose rate over an infinite flat plane contaminated with the isotope Cs-137. This can be done using published data (e.g. the USA EPA FGR12 Part 2). Results are in Table 1.

In 2010 I applied this method to the results of the study of Tondel et al 2004 of the effects of the Chernobyl fallout in northern Sweden in 1986. Tondel showed that there was an 11 percent increase in cancer rates in the ten years after Chernobyl for every $100kBqm^{-2}$ surface contamination. From Table 1 we see that $100kBqm^{-2}$ gives a dose rate of about 330nSv/h, an annual dose of 2.8mSv. Of course it is not the external dose that causes the cancer increase, it is the internal doses associated with that level of contamination, but we use the external dose rate as an indicator. The ECRR model, published in 2003, almost exactly predicted what these researchers found. The error factor relating the ICRP risk model was upwards of 600-fold. This was an external dose rate based on the contamination level that was employed by Tondel et al 2004, i.e. obtained from the Swedish authorities. The contamination was measured, not the external dose rate. In the case of Fukushima, we work backwards from the dose rate to the contamination. There are dose rate measurements reported, but also some surface contamination reports from the IAEA. I turn to the contamination and dose rates as reported in Japan up to 30th March.

Table 1 Contamination of surface based on gamma dose rate1 meter above ground (or gamma dose rate based on contamination of surface. Assumes photon energy of 660keV (Cs-137).

Gamma dose rate µSv/h	Surface contamination MBqm^{-2}
1	0.308
5	1.54
10	3.08
20	6.16
50	15.4
100	30.8
1000 (1mSv/h)	308
10,000 (10mSv/h)	3080 (3.08 GBq)

Table 2 Mean dose rates µSv/h and mean contamination (MBqm^{-2}) reported from 16th to 29th March 2011 at various distances from the nuclear site at Fukushima (Soma). SD is standard deviation; N is number of readings. Data from MEXT. (www.mext.jp)

Distances		Mean rate measured µSv/h	SD	N	MBqm^{-2} contamination deduced
0-20km	16/17 March	14.3	19.9	17	4.4
20-30km	16/17 March	11.9	18.8	39	3.7
30-50km	16/17 March	15.1	5.9	9	4.7
30-50km 29 March		6.42	9.7	18	1.9
50-70km 29 March		1.6	1.0	3	0.9

Radiation Exposure Near Fukushima

There were a number of sources of information, but for the purposes of the predictions I employ the official data from the Japanese Ministry of Education, Culture, Sports, Science and Technology (MEXT) and from the IAEA bulletins. The reported dose rates are given in Table 2. The measurements of surface contamination were reported by the IAEA in various bulletins. Their results are given in Table 3

Table 3 Surface beta gamma contamination, gamma dose rates in statements from 16th March to 29th March from the International Atomic Energy Agency, IAEA

IAEA bulletin date	Area/ contamination /dose	Statement
17 March	30km	In some locations at around 30 km from the Fukushima plant, the dose rates rose significantly in the last 24 hours (in one location from 80 to 170 microsievert per hour and in another from 26 to 95 microsievert per hour). But this was not the case at all locations at this distance from the plants. Dose rates to the northwest of the nuclear power plants, were observed in the range 3 to 170 microsievert per hour, with the higher levels observed around 30 km from the plant. Dose rates in other directions are in the 1 to 5 microsievert per hour range.
20 March	150km Tokyo	The IAEA radiation monitoring team took additional measurements yesterday between Tokyo and locations up to 150 km from the Fukushima site. Dose rates were typically a few microsieverts per hour compared to a typical background level of around 0.1 microsieverts per hour.
21 March	200km 2-160μSv/h 0.2-0.9MBq	As I reported yesterday, the IAEA radiation monitoring team took measurements at distances from 56 to 200 km from the Fukushima nuclear power plant. At two locations in Fukushima Prefecture gamma dose rate and beta-gamma contamination measurements have been repeated. These measurements showed high beta-gamma contamination levels. The dose-rate results ranged from 2-160 microsieverts per hour, which compares to a typical natural background level of around 0.1 microsieverts per hour. High levels of beta-gamma contamination have been measured between 16-58 km from the plant. Available results show contamination ranging from 0.2-0.9 MBq per square metre.

22 March	**68km** 0.8-9.1μSv/h 0.08-0.9MBq	The IAEA took measurements at additional locations between 35 to 68 km from the Fukushima plant. The dose-rate results ranged from 0.8 to 9.1 microsieverts per hour. The beta-gamma contamination measurements ranged from 0.08 to 0.9 MBq per square metre.
23 March	**30-73km** 0.2-6.9μSv/h 0.02-0.6MBq	The IAEA radiation monitoring team took additional measurements at distances from 30 to 73 km from the Fukushima nuclear power plant. Results from gamma dose-rate measurements in air ranged from 0.2 to 6.9 microsievert per hour. The beta-gamma contamination measurements ranged from 0.02 to 0.6 Megabecquerel per square metre.
24 March	**34-73km** 0.6-6.9μSv/h 0.04-0.4MBq **30-32km NW** 16-59μSv/h 3.8-4.9MBq	The IAEA radiation monitoring team made additional measurements at distances from 21 to 73 km from the Fukushima nuclear power plant. At distances between 34 and 73 km, in a westerly direction from the site, the dose rate ranged from 0.6 to 6.9 microsievert per hour. At the same locations, results of beta-gamma contamination measurements ranged from 0.04 to 0.4 Megabecquerel per square metre At distances between 30 and 32 kilometers from the Fukushima Nuclear Power Plant, in a north-westerly direction from the site, dose rates between 16 and 59 microsievert per hour were measured. At these locations, the results of beta-gamma contamination measurements ranged from 3.8 to 4.9 Megabecquerel per square metre. At a location of 21 km from the Fukushima site, where a dose rate of 115 microsieverts per hour was measured, the beta-gamma contamination level could not be determined.
27 March	**30-41km** 0.9-17μSv/h	The second team made additional measurements at distances of 30 to 41 km from the Fukushima nuclear power plant. At these locations, the dose rates ranged from 0.9 to 17 microsievert per hour. At the same locations, results of beta-gamma contamination measurements ranged from 0.03 to 3.1 Megabecquerel per square metre.

Comparisons of the IAEA statements and the MEXT measurements with our calculations of area contamination show that all the measurements and calculations roughly agree, although the contamination levels deduced from the dose rates are much higher than those reported by the IAEA. I will assume that the method can be used to determine the area contamination for those areas where no surface contamination levels have been reported. I employ the Tondel et al. 2004 published regression coefficient of 11 percent increase in cancer (all cancers) per 100kBqm^{-2} to predict the 10-year cancer yield. I assume for the calculation that the exposure is for 365 days. To establish the total cancer increase we will require the populations involved.

Using data from the Japanese census, it is possible to establish the approximate populations at risk. There are approximately 3.3 million people in the 100km radius and a further 7.8 million in the 100-200km band.

Exposure in Prefectures

Values of exposures in prefectures are also available from the MEXT Ministry website in the form of graphs of dose rate. These have been reduced to mean dose rates and are given in Table 4 for prefectures local to the catastrophe.

Table 4 Rates of exposure in selected prefectures from 16th to 29th March with deduced surface contamination. Background is assumed to be 0.04□Sv/h

Prefecture	Dose rate μSv/h	Deduced surface contamination kBq/m^{-2}
Ibaraki	0.35	95
Yamagata	0.1	18
Tochigi	0.2	50
Tokyo	0.1	18
Gunma	0.1	18
Saitama	0.1	18

Cancer Excess in 100km Population

Using an extremely conservative set of assumptions, I assume the 100km radius is contaminated uniformly to 600kBqm^{-2}. The dose associated with this level of contamination is 2μSv/h. Assuming that no one moves away and that the contamination remains at this level, using the Tondel et al. 2004 regression coefficient of 11 percent cancer increase per 100kBqm^{-2}, and assuming the same spectrum of radionuclides and pathways for exposure, the cancer increase in the 100km population is 66 percent and these cancers will be manifest in the next ten years.

The cancer rate in the Japanese population is 462 per 100,000 per year. Therefore the annual number of cancers in the 3,388,900 population of the 100km radius is 15,656. In ten years there will be 156,560 cancers normally if this rate is maintained plus an extra 66 percent of this number diagnosed from Fukushima that is 103,329 extra cancers due to the Fukushima exposures.

The ECRR absolute risk method cannot be formally used unless we know the individual radionuclide exposures. However it can be used if we approximate that 1/3 of the dose is internal and that 1/3 of the internal dose carries a weighting of 300 (which was the overall weighting factor obtained from the weapons test fallout spectrum of radionuclides epidemiology). Then the annual internal dose is 5.6mSv and 1/3 of this is 1.9mSv which we weight at 300. The total ECRR dose is thus 575mSvECRR. The collective dose is then 3,338,900 x 575 x 10^{-3} to give 1,919,867 person Sieverts and a lifetime (50 year) cancer yield of 191,986 extra cancers assuming the ECRR risk factor of 0.1 per Sievert ECRR. Given the different time frames, these numbers obtained from the Tondel et al. 2004 regression and the ECRR absolute model based on the atmospheric test cancer yields in Wales and England are in reasonable agreement. The three predictions are given in Table 5

Table 5 The predicted cancer increases in the 100km zone near the Fukushima site

Model	Cancer yield	Note, assumptions
ICRP	2,838	In 50 years, based on collective doses at exposure of 2 µSv/h for one year
ECRR Tondel	103,329	In ten years following the catastrophe, based on surface contamination only
ECRR absolute	191,986	In 50 years, based on collective doses at exposure of 2 µSv/h for one year; probably half of these expressed in the first ten years.

Cancer Excess in 200km Annulus Population

The methods employed above may be extended to the 200km annulus if the contamination levels are known. Presently no data is available of contamination in these areas although dose rates are available. NOAA computer modelling carried out by us and published on the internet and elsewhere suggest that the plumes from the catastrophe have travelled south over the highly populated areas in and near Tokyo. Dose rates have been published for these areas and from these dose rates it can be assumed that significant exposures have occurred. From Table 4 and Fig. 3 we can assume that the exposures are of the order of 1 µSv/h with associated contamination levels. Therefore the methods employed for the 100km area may be extended to the 200km area. The population is, however much greater at 7,874,600. The results are given in Table 6.

Table 6 The predicted cancer increases in the 100-200km zone near the Fukushima site

Model	Cancer yield	Note, assumptions
ICRP	3,320	In 50 years, based on collective doses at exposure of 2 µSv/h for one year
ECRR Tondel	120,894	In ten years following the catastrophe, based on surface contamination only
ECRR absolute	224,623	In 50 years, based on collective doses at exposure of 2 µSv/h for one year; probably half of these expressed in the first ten years.

Testing the ICRP Approach

We can compare the ICRP model. The annual dose from this contamination can be calculated in mSv. If we assume 365 days and 24 hour a day exposures then for 2 µSv/h the annual dose is 17mSv. The population is 3,338,900 so the collective dose is 56,761 person Sieverts. The ICRP absolute cancer risk factor is 0.05 per Sievert. Thus the ICRP predicts 2838 extra cancers in this population from the Fukushima fallout. Given that the ICRP predicted excess cancers will probably appear in the next 10 years, they will not be measurable above the normal rate unless they are rare cancers. Examples are leukemia or thyroid cancer in children. The latter has already been reported, demonstrating the complete failure of the ICRP model which predicts no increases. In fact, there have been 33 confirmed thyroid cancer cases in the age group 0-18 in those screened in Fukushima Prefecture.

Thyroid Cancer in Young People

It was reported recently (December 21, 2013) that a survey of thyroid conditions in young people age 0-18 by Fukushima Medical University have now found 33 confirmed and 41 possible cases of thyroid cancer in 269,354 individuals screened. This is in a (almost) three-year period. The 2005 Japanese national incidence rate for thyroid cancer aged 0-18 is given in a recent peer reviewed report as 0.1 per 100,000. That means in the last three years we would expect 0.8 cases: we actually see 33 (or a possible maximum of 74), giving a Relative Risk of 41 (or up to 91). *The Japan Times* tells us "Researchers at Fukushima Medical University, which has been taking the leading role in the study, have said they do not believe the cases are related to the nuclear crisis."

The UNSCEAR would agree. Also the WHO and the IAEA. In its preliminary report on Fukushima health effects, issued in 2012, UNSCEAR stated that the maximum thyroid dose was 35mSv and that most received a lot less. If we assume that the mean thyroid dose was 20mSv, the predictions of the latest ICRP model in a population of 200,000 is 13 cases in 50 years.

Other Areas and Some Caveats

The ECRR risk model assumed a biphasic dose response relationship (up, then down, then up again as the dose increases) and so there is linearity of risk only over the low dose region. For the high exposures modeled here there will be significant saturation, that is, the proportionate cancer risk will fall as the dose increases. This is for a number of reasons but partly because there will be competing causes of early death. We can examine the competing causes of death using data from the Chernobyl effects but first I will look at Tokyo. Tokyo has by far the largest population at risk, and though the distances are greater, there was significant contamination as I know from having personally made measurements.

Tokyo: Using Car Engine Air Filters.

From the beginning, my colleague Dai Williams and I were employing the US NOAA air modeling system (HYSPLIT) to track the releases from Fukushima. Although the early releases were blown out to sea, later the wind direction changed and the radionuclide cloud was swept back over Japan just north of, but crossing, Tokyo, and then moved north over the eastern part of Japan. According to measurements made on the roof of the US Embassy in Tokyo, which were leaked to me, the major increases in contamination appeared in Tokyo on the Wednesday and Thursday a week after the explosions and after the fire in the Reactor 4 building. The US Embassy staff sent a team out to the north to obtain air samples from 19th March, but were also routinely sampling and measuring on the Tokyo Embassy Roof and the Harris Tower. These were analyzed for the different fission radionuclides. The high levels of radioactivity discovered in Tokyo air were not, of course, communicated with the people who were inhaling the particles, and the media reports continued throughout the period to provide reassurance that there was no exposure as far south as Tokyo.

Remember, the first explosion of Unit 1 occurred on the 12th March. Then, on 14th, Unit 3 exploded, followed by Units 2 and 4 on the 15th. But something unreported went wrong on Monday 21st, a week after the explosions. For example, on the Embassy Roof on 19th March at 12:50 local time, the leaked US DoE EXCEL file that I have listed 12,580mBq/m^3 total beta and 42.1mBq/m^3 alpha, with 275mBq of I-131. At 11:35 on 20th there were 23,865mBq beta and 555mBq alpha with 103mBq I-131. At 12:45 on 21st the beta counts were 6,031,000mBq/m^3 beta and 1,302mBq alpha with 1,039mBq I-131 and a whole list of other radionuclides including Tellurium-132, Caesium-137 and -134, Niobium-95, and Tellurium-129m.

When I visited Japan in June at the invitation of Mr Yanagihara and the lawyers who were calling for the evacuation of the children from the contaminated areas I asked people to obtain vehicle air filters and post them to me. I received about 30 filters with details of the car and how long it had been driven after March 11th before the filter was removed. Car engines breathe air like people, and indeed it is possible to calculate the amount of air that is sucked through a car engine filter on the basis of the engine size and how many miles the car is driven (which I asked them to attach to the filter). I measured Caesium-137 and Caesium-134 in at least 20 car filters that were sent, including some from Tokyo and Chiba. They all contained radioactivity, some were so radioactive that when I sent them for high resolution gamma spectrometry, the lab I sent them to complained that it was illegal to post them though the UK mailing system. A low resolution gamma spectrum of a car engine filter from Tokyo is shown in Fig. 2 where the Caesium-134 peaks are identified. An autoradiograph of a car filter is shown in Fig. 3 below (obtained by Marco Kaltofen).

The levels of Caesium-137 in the car filter dust in the filters I examined were as high as 100,000 Bq/kg, but these filters are quite coarse and do not trap the very small particles less than 10 microns, so the air concentration must have been at least twice this. I was able to calculate that levels of Caesium-137 radioactivity in air in Chiba and Tokyo were as high as $1000mBq/m^3$, and later the DoE data in the EXCEL file showed that this was probably a significant underestimate. It is also clear that there were alpha emitters in the mix. We can compare these results with weapons fallout and Chernobyl. In the UK, measurements of Cs-137 made at Harwell, England at the time of the peak in global weapons fallout from atmospheric testing in 1962/63 showed the level to be $2.7mBq/m^3$ ($0.0027Bq/m^3$). The peak after Chernobyl was about $8mBq/m^3$. After the Chernobyl fallout had faded in 1990 the Cs-137 content of air was below $0.001mBq/m^3$. The value in Fukushima give by the air filter analysis is $1,000mBq/m^3$ but probably more than 10 times this value given the inefficiency of the filter to trap the sub micron particles. It should be pointed out that the weapons testing levels caused a 2 percent increase in infant mortality at the time and a 30 percent increase in cancer some 20 years after the fallout.

Figure 2 Low resolution gamma spectrum of a car filter from Tokyo (Nissan Cube) showing Cs-134 spectrum superimposed.

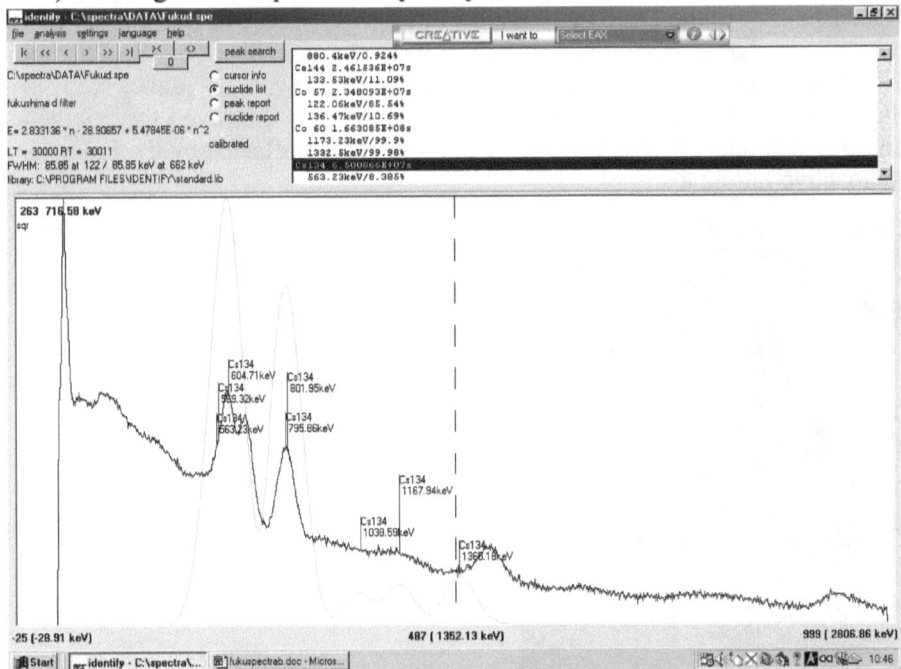

The most worrying measurement I made was of an air conditioning unit filter that was sent to me by a lady who lived on the 20th floor of an apartment block near the Tokyo Tower. This showed about 60,000Bq/kg of Caesium and also high levels of Uranium-235 and Lead-210. But unlike the car filters, this unit was located <u>inside</u> the apartment and merely recycled air. They left Tokyo and switched off the unit on the Tuesday 15th March, and did not return for some months, so all this contamination was inside their apartment between the 12th and 15th!

Non Cancer Health Effects: The End of the Population of Northern Japan?

The calculations made above are for cancer. But we know from Chernobyl that these internal exposures cause an increase in a wide range of illnesses, including a huge increase in deaths from stroke and heart attack even in children. In fact all this is already happening if you listen to the jungle telegraph. Studies of populations exposed internally show that a wide spectrum of diseases and conditions follow; these include heart disease, diabetes, and all the normal conditions and illnesses that contribute to mortality and morbidity. In addition, epidemiology of nuclear test veterans, Chernobyl-affected populations and those exposed to Uranium together show us that alarming increases in congenital disease in children and grandchildren are to be expected. Chernobyl has taught us

91

that the overall effects of the Fukushima contamination on the populations exposed will be catastrophic. In Fig. 4, I copy data presented by Dr Yuri Bandashevsky to the ECRR conference in Lesvos, Greece in 2009. It shows the remarkable changes in birth rate and death rate brought about by the Chernobyl effects in Belarus. The result is that the population dynamic cannot recover as there is now insufficient replacement. Chernobyl has destroyed the population and this will happen with those parts of North Japan affected by the Fukushima radioactive fallout also.

Figure 3 Autoradiograph of a car filter from Tokyo showing radioactive particles

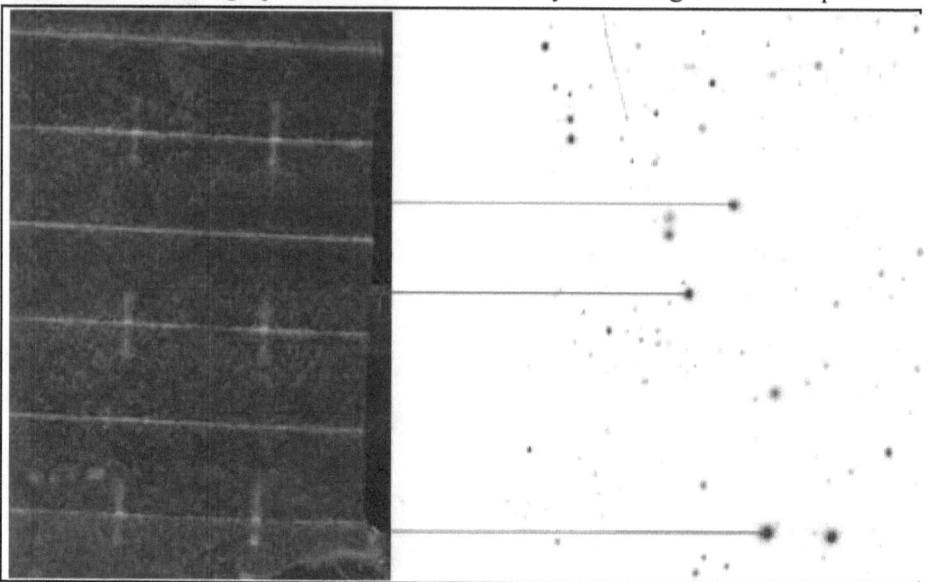

Source: Marco Kaltofen, *High Radioactivity Particles in Japanese House Dusts,* Nuclear Science and Engineering, Worcester Polytechnic Institute, March 2014.

Figure 4 Variation in the birth and death rates per 1000 inhabitants of Belarus from 1950-2003. Note the sudden increase in death rate and decrease in birth rate after Chernobyl.

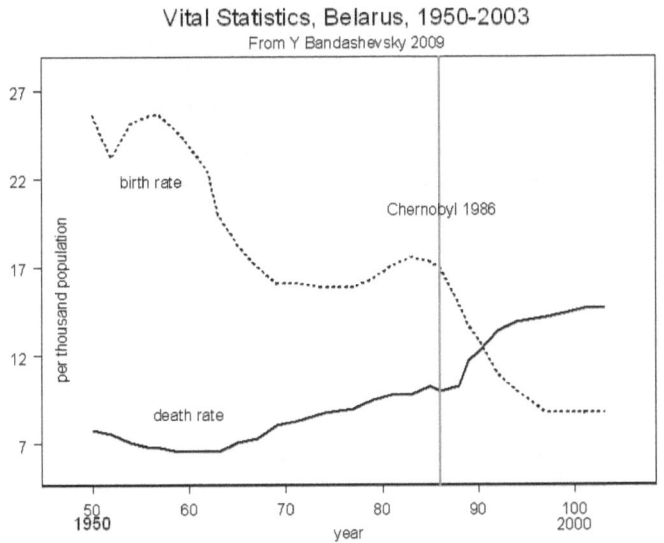

Effects Further Away

The releases of radioactivity from the catastrophe included fairly short lived gases like the radio Xenons and Iodines, volatile particulates including fuel Uranium and Plutonium particles which were contaminated with short-lived nuclides and longer lived Caesium-137, Caesium-134, Cerium-144 and Strontium-90. The total quantities are now known from published measurements by those tracking the plumes in various countries. The initial releases, comparable for some nuclides, are considerably greater than the releases from Chernobyl. This is not surprising since at Fukushima there were four reactors and five spent fuel pools involved compared with one reactor at Chernobyl. And, of course, the affair continues. The molten fuel is out of containment and being continuously cooled by water pumped into the reactor buildings. This water dissolves the radioactive elements in the hot masses of melted fuel and contaminates the groundwater. Because the reactors are close to the ocean, the radioactive water reaches the ocean, although some of it is pumped out of wells and stored in huge tanks. No-one knows what to do with this contaminated water in the tanks and new tanks have to be built to take the radioactive water which continues endlessly to appear.

The concentrations in the air of radionuclides from the initial explosions were measured by various laboratories around the world. The highest levels reported were from the West Coast of the USA. But these levels of fallout were not anywhere near as high as levels which had been obtained during the atmospheric tests, especially since many of these tests had been in the Nevada and New Mexico test sites. Of course, these USA tests had caused increases in cancer and leukemia and also thyroid problems, but the fear of Fukushima seemed to be pathological. Everyone was buying Geiger counters (and that is a good thing, in my opinion) to see if the US government was lying about the levels of fallout from Fukushima. But the effects of the fallout seemed to be detectable. One set of researchers later looked at infant mortality on the West Coast and discovered a 35 percent increase associated with Fukushima. This naturally created a furor, but follow ups using more data seemed to confirm this as a real effect. My own involvement was to work with Joe Mangano and Janette Sherman on a study of newborn hyperthyroidism and Fukushima radio-Iodine in California. This also seemed to show a significant correlation. So although the levels of radioactivity getting to the USA were tiny, compared with earlier exposures from weapons fallout and even, of course, from local nuclear power station releases (of which there are many in the USA), the responses to these "low doses" were very real.

The radionuclides from the initial explosions mostly contaminated the Pacific. The ongoing releases are also to the ocean. This seems to be having a catastrophic effect on the ocean life. A large number of published studies and observations point to a post-Fukushima collapse of life in the northern Pacific. This includes sudden loss of whole species (sardines, starfish), increases in fatal diseases in sea mammals, a whole range of strange and extraordinary effects which taken together point to a disaster of the first magnitude. It is clear that the massive contamination of the Pacific is not dispersing uniformly, but rather is drifting towards the coast of the USA, and according to calculations will fetch up on the beaches of California sometime this year or next (2015). It is not there yet, or at least not measurably so. When it does start to contaminate the beaches, there is a new concern that no one has drawn attention to. This is sea-to-land transfer.

Sea to Land Transfer

There is a frightening and virtually unknown consequence of releasing radioactivity to the sea. This was discovered in the 1980s when the releases to the Irish Sea from the Sellafield plutonium reprocessing plant began to appear on the coast of the UK and Ireland. The investigators from the UK Atomic Energy Research Establishment near Oxford found that the action of the waves caused radioactive particles which appeared on the coasts (rather than being somehow diluted in the sea, as had been predicted) to become thrown into the air and drift ashore, where those who lived within about 1km of the sea coast inhaled them. The effects of this were examined in a huge epidemiological study of the Welsh population which I carried out in 1998-2001 funded by the Irish State in connection with a court case. There was a sharp increase in cancer very near the contaminated coasts which fell off rapidly as you moved inland. The effect was there for all cancers but most clear in the children. In a follow up study I published in 2004 there were childhood leukemia rates more than 12 times those expected based on national population. This effect had been already found near Sellafield itself in the 1980s, but the response by the authorities on the basis of the ICRP model was that the "doses" were too low. The first point is that the releases to the ocean in Japan will fetch up all along the east coast of Northern Japan, so that those towns and villages which are on the coast, no matter that they are hundreds of miles from Fukushima, will suffer a high level of risk from this effect. It is also possible, indeed likely that this process will also occur on the West Coast of the USA, though we will have to await the arrival of the contaminated surface water to see what the magnitude of the contamination is. Presumably there will also be a problem anywhere else this material fetches up, in China, Korea etc. so those people also need to watch out.

Figure 5 Alpha particle tracks in CR39 plastic from hot particles in an edible mussel from Ravenglass near the Sellafield plant in the UK.

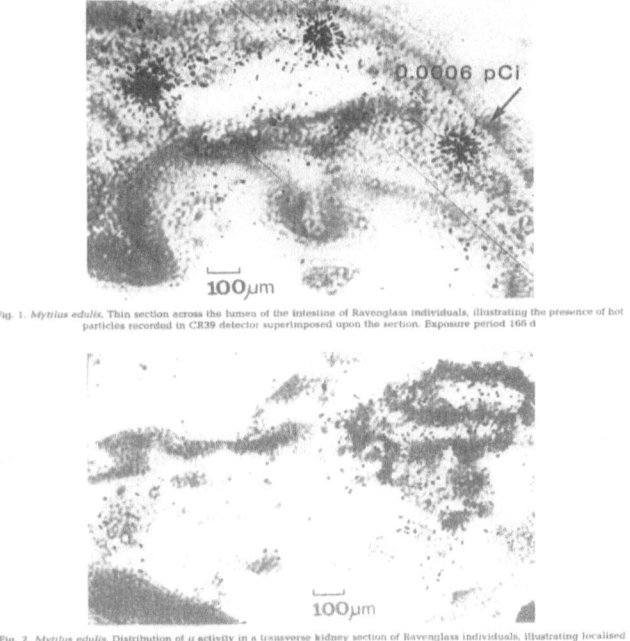

Fig. 1. *Mytilus edulis*. Thin section across the lumen of the intestine of Ravenglass individuals, illustrating the presence of hot particles recorded in CR39 detector superimposed upon the section. Exposure period 166 d

Fig. 2. *Mytilus edulis*. Distribution of α activity in a transverse kidney section of Ravenglass individuals, illustrating localised enriched levels of activity. Exposure period 166 d. Procedures as in Figure 1

Captions:
Fig. 1. *Mytilus edulis*. This section across the lumen of the intestine of Ravenglass individuals, illustrating the presence of hot particles recorded in CR39 detector superimposed upon this section. Exposure period 166 d
Fig. 2. *Mytilus edulis*. Distribution of α activity in a transverse kidney section of Ravenglass individuals, illustrating localised enriched levels of activity. Exposure period 166 d. Procedures as in Figure 1

I mentioned earlier that releases contained micron and sub micron diameter particles. The alpha emitters Uranium and Plutonium are released in the form of insoluble alpha emitting particles, which are found in air filters (as we saw with the results from the Tokyo filters). But these particles also end up in the sea creatures on the contaminated coasts, and of course in those who eat them. The Japanese diet has a high level of sea food, shellfish and fish. In Fig. 5 I show an image of the alpha tracks from particles contaminating an edible mussel from the Irish Sea near the Sellafield nuclear fuel reprocessing plant.

Worst Case Scenario

Because of the way in which ocean currents and atmospheric air movements operate, I do not think that there will be any measurable effect in the Southern hemisphere, and the concentrations in Europe of the Fukushima material (which we monitored and even measured) were too low to cause any effects. Unlike the atmospheric test fallout, Fukushima is mostly a local affair. But what if it totally blew up? What if some collapse of a building or some inadvert unfortunate manipulation of the fuel rods that are currently being removed from the Reactor 4 building resulted in a nuclear explosion that involved all the radioactive material, everything? Well I have looked at this "worst case scenario" also. Many have predicted "the end of life on earth" if this should occur or have advised everyone in the USA to emigrate to Australia or Argentina. But it is easy to show that this is nonsense. Although, in a way, it would add to and be part of an ongoing "end of life on earth" scenario that began in 1945 and shows little sign of stopping, the total amount of radioactivity in all of the Fukushima reactors combined is about equal to the total releases to the planet from all weapons tests. And although those gave us the cancer epidemic, the loss of fertility, low sperm counts and had a generally frightful effect on "life on earth", they did not precipitate the "end of life on earth". And those nuclear test releases were global. There are even people still alive and kicking at the Marshall Islands, though the radiation profoundly affected the population there. The final act of the tests, the 1962 Soviet Tsar Bomba of 40 Megatons, injected about 1mSv worth of Strontium-90 into the Northern hemisphere and it is still raining down today. Nothing like that can happen in Fukushima. There will not be a megaton thermonuclear fusion bomb effect even if everything went up. It would be largely a local affair. Not good, however, I should add, and would certainly put northern Japan into "Mad Max" territory.

How to Protect Yourself, What to Do

The risk model of the ECRR associates one major radiation risk from internal exposure in the high DNA doses from those elements of Group 2 in the Periodic Table that have a high affinity for DNA. This effectively means Strontium-90, Barium-140, and the Uranium isotopes 238, 235 and 234, all of which were components of the Fukushima releases. There were concerns during the period of the weapons test fallout also about Strontium-90, although the Uranium exposures were not then believed (for some reason) to be significant. As a result of the use of Depleted Uranium in Iraq and the Balkans, and the terrible cost in cancer and birth defects that this has caused, we now know better. The genetic effects of Strontium-90 were investigated in experiments with mice carried out in Sweden in 1963. The results showed that exposure of male mice to Sr-90 followed by mating caused foetal death in the offspring and effects to the second generation. Russian studies demonstrated that the deaths were a consequence mainly of heart and circulatory system defects. However, a number of experiments were carried out to see if taking non-radioactive Strontium or even Calcium could protect against the binding of the radioactive Strontium-90 to the DNA. These experiments (and there were several papers published) showed that the method worked.

Shortly after Fukushima occurred I proposed on TV and in the media to the population of Japan that they should evacuate from the area up to 100 miles from the nuclear site. If they could not, I advised that they should take Calcium tablets to block the access of Strontium and Uranium to the DNA. A number of organizations began to market Calcium tablets for radiation protection. One company asked me if I would allow them to use my name to market the Calcium tablets. I formulated the tablet and agreed. I was then attacked in the newspapers in the UK for making money out of the suffering of the Japanese people, and for selling overpriced and useless tablets. I was pleased to see that many people took up my advice: I am sure that many deaths have been prevented. But the attacks in the UK *Guardian* newspaper had a significant effect on my life. The newspaper would not allow me to respond, and the Press Complaints Commission also would not allow me to respond. The newspaper quoted a Dr Gerry Thomas as stating that radioactive material did not bind to DNA and that Calcium tablets were useless. Thus the scientifically proven and published evidence could be overridden in the national media by someone who, as it turned out, had no research publications in the area of radiation and health whatever.

Concluding Thoughts

I raise this issue here because it is part of the general question of science as truth, the question of dispossession of truth by power. Where we came in. Power nowadays is with the media, and can be, and is manipulated. The whole Fukushima affair, which is by no means over, is characterized by the kind of disinformation and airbrushing that in the 1950s was associated with the Soviet Union. Of course, maybe even then the same thing was happening over in the West. In fact, we now know it was. The massive increase in heart disease, blamed upon diet, was actually caused by increased dairy product intake, and the dairy was full of Strontium-90 and Uranium-238 from the bombs. The heart attacks began in (and rates were highest in) those areas where the fallout was greatest, areas of high rainfall. The increase in lung cancer from cigarette smoking was caused by the increased quantities of radionuclides in the fallout which was incorporated into the tobacco. Prof Ed Radford, who began to question the smoking and lung cancer issue was rapidly removed from the BEIR III committee. The "thalidomide baby" genetic birth defects may, according to a study carried out in Canada, have been caused by the fallout particles, just as the Uranium particles in Fallujah Iraq caused the same spectrum of births defects there. No one identified the infant mortality cause in the 1960s, indeed there was a confidential enquiry in England, but no cause was found.

In the UK now for some years there has been a media operation called the "Science Media Centre". This outfit, which is funded by big business and nuclear operators (e.g. Electricite de France, Monsanto) have a whole bagful of tame scientists that they argue give the media the correct take on any story that may suggest to the public that they are being systematically poisoned. No doubt there is a similar operation in the USA. And it is not just defensive operations. There is clearly a team of people who systematically attack those scientists, like myself, who draw attention to the effects of radiation exposures. I have been thrown out of three Universities because of the letters and attacks that have been made against me and my work, the University of Liverpool, the University of Ulster and most recently Jacobs University, Bremen. I have lost all my funding from the Charities that used to support my work. I have had scientific papers rejected without even being sent to reviewers following letters written to the journal editors, on occasion even before I have sent the paper to the journal.

There is no doubt that the health data, the information about increases in ill health in those exposed to the Fukushima radioactivity, are being controlled. There is one easy proof. The population exposed in the 200km radius is about 10 million. That would include about 1.4 million children aged 0-14. The background rate of childhood leukemia is about 6 per 100,000 per year. Therefore in the three years since Fukushima we would expect, normally, if nothing had happened, if there had been no radiation, 85 cases of leukemia per year or a total of 255 cases. Now if you are a mother, and your child developed leukemia after Fukushima, would you not want to scream from the rooftops? I mean, you would not know that the rate had gone up or not. You would put something on Youtube, at least one of the mothers or father would wouldn't they? Would the Japanese newspapers not want to draw attention to your child's illness? Have a whip-round to send her or him to Disneyland, like they do here in Wales. Of course. But not one cheep out of anyone. Nothing.

So the message is plain. The message of Fukushima, because it was such a terminally dangerous event for the nuclear and military complex, and all the money riding on the Uranium shares and energy futures, had to be controlled through a massive exercise of power. And such an exercise of power cannot be hidden. It is there for all to see from the nonexistent leukemia rates to the absurd explanations about the thyroid cancers being detected. But that does not mean that no one can do anything about it except look after themselves and their loved ones. If there is one message from Fukushima it is this: don't believe the scientists working for the governments. Oh, and get as far away as possible.

References

Busby C.C. (2009) Very Low Dose Fetal Exposure to Chernobyl Contamination Resulted in Increases in Infant Leukemia in Europe and Raises Questions about Current Radiation Risk Models. *International Journal of Environmental Research and Public Health.*; 6(12):3105-3114. http://www.mdpi.com/1660-4601/6/12/3105

Busby Christopher (2013). Aspects of DNA Damage from Internal Radionuclides, New Research Directions in DNA Repair, Prof. Clark Chen (Ed.), ISBN: 978-953-51-1114-6, InTech, DOI: 10.5772/53942. Available from: http://www.intechopen.com/books/new-research-directions-in-dna-repair/aspects-of-dna-damage-from-internal-radionuclides

Busby C, Busby J, Rietuma D and de Messieres M—Eds (2011) Fukushima—what to expect. Proceedings of the 3rd International Conference of the European Committee on Radiation Risk May 5/6th 2009 Lesvos Greece Aberystwyth: Green Audit

ECRR2010 The 2010 Recommendations of the European Committee on Radiation Risk. Edited by Chris Busby, Rosalie Bertell, Alexey Yablokov, Inge Schmitz Feuerhake and Molly Scott Cato. Brussels: ECRR; available from www.euradcom.org

Kaltofen, Marco, (March 2014) *High Radioactivity Particles in Japanese House Dusts,* Nuclear Science and Engineering, Worcester Polytechnic Institute. Available: http://bostonchemicaldata.com/wpi/mKaltofenNagoya2014.pdf

The Lesvos Declaration (2009) see www.euradcom.org

Ministry of Education, Culture, Sports, Science and Technology (MEXT) (www.mext.go.jp/english/radioactivity_level/detail/1303986.htm)

Tondel Martin, Lindgren Peter, Hjalmarsson Peter, Hardell Lennart and Persson Bodil, (2006) Increased incidence of malignancies in Sweden after the Chernobyl accident, *American Journal of Industrial Medicine*, (49), 3, 159-168.

Tondel M, Hjalmarsson P, Hardell L, Carisson G and Axelson A (2004) Increase in regional total cancer incidence in Northern Sweden. *J Epidemiol. Community Health.* 58 1011-10

Fukushima and Dispossession: The End of Liberal Democracy in Japan?

Majia Nadesan

Japan's former Prime Minister, Naoto Kan (2013), described the Fukushima Daiichi nuclear disaster as "the most severe accident in the history of mankind." Although precipitated by natural events, the reactor explosions and subsequent crisis management were human engineered fiascos. This summary conclusion was reached by the National Japanese Diet, who declared in their 2012 official report, "The Fukushima Nuclear Accident Independent Investigation Commission" that human error was, above all else, responsible for the disaster. The report's chairman, Kiyoshi Kurokawa, introduces findings with these words: "Our report catalogues a multitude of errors and wilful negligence that left the Fukushima plant unprepared for the events of March 2011. And it examines serious deficiencies in the response to the accident by TEPCO, regulators and the government."

Yet, despite the Fukushima catastrophe, Japan's government, especially the Liberal Democratic Party (LDP) leadership, continues to push nuclear energy forward (Nagata, 2014). In 2014, Prime Minister Abe of the LDP party endorsed nuclear power as an "important base load power source" for national stability, while pushing for reactor restarts, pending approval by Japan's Nuclear Regulation Authority (NRA). Public misgivings about reactor safety in earthquake-prone Japan are growing ("Nuclear Plant Restarts," 2014). An academic report provides grounds for concern through its description of damage incurred at fourteen nuclear reactors in Japan caused by the March 2011 earthquake (Ohmae, 2011). These reactors may be especially vulnerable to irreparable structural damage in future earthquakes, especially since several reactors have been found to be situated directly on faults.

The push for nuclear power in Japan, the US, and the UK (among other places) defies reason. The problem of long-term nuclear waste management alone belies the promise of an atomic utopia. Moreover, conditions at Fukushima Daiichi demonstrate conclusively that nuclear disasters pose catastrophic risks for property and personhood (see Nadesan, 2013). Today, the Fukushima nuclear disaster remains unmitigated, continuing to "leak" contamination into the aquifer, the ocean, and the atmosphere. In the fall of 2013, a TEPCO official described the Fukushima Daiichi plant as "out of control" ("TEPCO Official Says," 2013). TEPCO denied the claim, although acknowledging uncertainty about the conditions and location of fuel formerly in Fukushima Daiichi reactors 1 through 4 because radiation levels preclude investigation (Kimura & Hattori, 2014). TEPCO's official plan for decommissioning Daiichi calls for fuel removal to begin at reactors 1 and 2 in 2020, but an anonymous TEPCO official told the Australian Broadcasting Corporation in March 2014 that the reactors could never

100

be de-contaminated because the technology for extracting melted fuel does not exist (Carney, 2014). Meanwhile, the site continues to contaminate the ocean, fresh water aquifer, and atmosphere.

This chapter argues that nuclear power produces many externalities that are fundamentally incongruous with liberal democracy itself. The Organization for Economic Co-operation and Development (OECD, 2001) defines environmental externalities as "uncompensated environmental effects of production and consumption that affect consumer utility and enterprise cost outside the market mechanism." Environmental externalities shift costs from polluters to bystanders, including private citizens and future generations. In this fashion, externalities distort market pricing and force impacted bystanders to absorb full production costs. This chapter considers this argument by documenting externalities of the Fukushima disaster, while discussing their implications for the commons, citizen rights, and democracy itself as Japanese citizens are dispossessed of rights of personhood, property, and expression. Liberal democratic capitalism is itself at risk from nuclear power.

The Disaster of the Commons

The Fukushima disaster is above all else a disaster of the commons. The March 11, 2011 Richter scale 9 earthquake that rocked Japan's north-east coast caused unprecedented damage to nineteen nuclear reactors in Japan. The Fukushima complex suffered the worst reported damage, leading to significant atmospheric plumes of radiation and contamination of ground and sea water. Three years after the March 11 earthquake, contaminated water from the site continues to flow freely into the ocean. Fresh and sea water contamination are rising. Technology for removing the most highly radioactive debris has not yet been invented.

The Fukushima complex, located 160 miles north of Tokyo, is composed of the Daiichi and Daini sites, located approximately six miles apart. The Daiichi site, which sustained the most damage, has six reactors and seven spent fuel pools, the latter containing 1,760 tons of spent fuel inventory as of 2010 (Kumano, 2010). Reactors 1 through 3 operational at Fukushima Daiichi at the time of the earthquake all suffered severe damage, including explosions, and melt-throughs (Institute of Nuclear Power Operations, 2011; "Government Report to IAEA," 2011; McCurry, 2011, Ryall, 2011). Reactor 4, which was not operating at the time of the disaster, was also damaged extensively, and fires in its spent fuel pool were reported to the IAEA on 14 March and 15 March.[i] Fire was reportedly extinguished later on 15 March.[ii] *The Los Angeles Times* reported this same sequence of fires in unit 4 spent fuel pool, which purportedly contained both new and spent fuel (Hall & Williams, 2011).

Radiation levels at the site continue to rise three years after the March 2011 explosions. TEPCO reported in December 2013 that a vent tower was 25 Sieverts an hour, which is a level of exposure that kills humans very quickly ("Record

Outdoor," 2013). TEPCO persists in denying knowledge of reactor fuel location (Nishikawa, 2014). Speculation exists that some of the melted fuel may reside uncontained in the earth. Perhaps it is cooled by the underground river, which was returned to its historical course by the earthquake (Nagata, 2013). The river was diverted when Daiichi was built decades ago. The river runs at about 1,000 tons daily, with TEPCO announcing that approximately 400 tons of that penetrates reactor buildings 1 – 4. Water saturation is contributing to ground liquefaction, which poses direct risks to the reactor buildings and common spent fuel pool. Contaminated ground water is also flowing into the ocean (Nagata, 2014).

Ground water contamination is extraordinarily high, with water from the well between the ocean and unit 1 measuring a record 5 million Becquerels per liter of radioactive Strontium-90 alone in July 2013 ("Record Strontium-90 Level," 2014). The quantity of Strontium-90 in the July 2013 well water was 170,000 times the permissible level. TEPCO stated the *total* Becquerels per liter was likely 10 million when all beta ray sources are included ("TEPCO to Review," 2014). TEPCO had originally interpreted the July 2013 beta tests as indicating 700,000 Becquerel per liter of Strontium-90, but revised the figure upwards to 5 million in February of 2014. Critics charged that TEPCO was hiding data after the company announced it would not revise known inaccuracies in data reported from July 2013 through December 2013 ("TEPCO Withheld," 2014). The results that had been reported by TEPCO during this period and after indicated a *clear upward trend in strontium, tritium and cesium contamination levels from July 2013 through February of 2014* ("TEPCO Announced," 2014).[iii] It appears likely that actual levels of contamination in the spring of 2014 are in excess of the revised figure of 10 million Becquerels per liter.

Unprecedented volumes of highly radioactive water are produced at Daiichi through contamination of ground water and from TEPCO's ongoing injections aimed at keeping cool fuel remaining in reactors and spent fuel pools. TEPCO has limited capacities for capturing and containing the contaminated water. TEPCO's filtration systems cannot remove tritium and strontium. Water leaking from storage containers has measured up to 230 million Becquerels per liter as of February 2014 ("TEPCO Finds," 2014). Storage of water with very high beta readings at the site is reaching capacity at over 360,000 tons in storage as of February 2014 (Varma, 2014). TEPCO announced at a January 2014 press conference that contaminated water in storage tanks is now producing Bremsstrahlung radiation, which contributes to rising atmospheric radiation levels at the Daiichi site (Mochuzuki, 2014). TEPCO maintains that contaminated water must be dumped into the ocean, a position that is also supported by the IAEA. The levels of fresh and sea water contamination have been rising since mid-2013, for reasons that will be discussed presently.

The March 2011 explosions ejected highly contaminated debris and plumes of smoke and steam that rose high into the atmosphere. For example, debris found at the river mouth in Naraha-machi 15 kilometers from the Daiichi site measured

2.92 million Becquerels of Cesium-137 (alone) per 0.4 gram sample (Ex-SKF, 2014). The IAEA defines land contamination at 40 kiloBecquerels per meter squared, which is 40,000 Becquerels (Christoudias & Lelieveld, 2012). The Naraha-machi river mouth sample was very hot in comparison, for cesium alone. Contaminated fresh water poses direct risks to Fukushima agriculture. In March of 2014, *The Asahi Shimbun* reported that soil samples taken from 1,939 reservoirs in Fukushima Prefecture contained high levels of Cesium, exceeding 100,000 Becquerels per kilogram in 14 reservoirs, and exceeding 8,000 Becquerels per kilogram in 576 reservoirs (Fujiwara, 2014). These levels suggest significant contamination.

Unfortunately, very few empirical estimates exist on overall levels of fallout and contamination throughout Japan. Most research on Fukushima fallout patterns has modeled dispersion using TEPCO-provided source terms for their projections of the distribution of radionuclides through wind and ocean currents. These models are useful for understanding the likely directionality of fallout plumes, but do not lend insight into total contamination levels, as noted by European scientists who used TEPCO source terms to estimate contamination zones in Japan. Using the IAEA definition of contamination, the scientists concluded that the contaminated zone ranged in area between 34,000 and 56,000 kilometers squared (Christoudias & Lelieveld, 2012). However, source terms for total contamination offered by TEPCO have been revised upwards many times. TEPCO stated and re-stated atmospheric emissions from March of 2011 in reports ranging from a low of 440,000 teraBecquerels to a reported high of 900,000 teraBecquerels, which is "900 quadrillion Becquerels," or "17 zeroes (a quadrillion is one thousand trillion)" ("TEPCO Post-Mortem," 2012). Japanese critics challenge TEPCO's emission calculations. For example, Professor Hiroaki Koide, of the Kyoto University Reactor Research Institute has suggested that the Japanese government's report underestimated the amount of Cesium-137 released in the atmosphere. He estimated atmospheric Cesium-137 contamination from the initial explosions was up to 500 times greater than produced by the Hiroshima bomb (cited in De Alou, 2013).

Limited field sampling challenges estimations of overall fallout. Most samples of contamination resulting from the disaster were collected in 2011 and, at latest, 2012. One study released in *Scientific Reports* published by *Nature* titled "Isotopic evidence of plutonium release into the environment from the Fukushima DNPP accident" by Zheng et al. (2012) found that a wide array of highly volatile fission products were released, including ^{129m}Te, ^{131}I, ^{134}Cs, ^{136}Cs and ^{137}Cs, which were all found to be "widely distributed in Fukushima and its adjacent prefectures in eastern Japan." The study also called for long-term study of actinides, after finding high activity ratios of plutonium isotopes. In a separate study, soil sample data collected at the Canadian Embassy in Tokyo on March 23, 2011 were used to extrapolate total fallout inventory at 225,000 Becquerels per square meter (Zhang, Friese, & Ungar, 2013). Atmospheric fallout was not restricted to Japan. A 2012 U.S. Geological Survey documents wet deposition of fission products in samples collected in 2011 in the western US (Wetherbee, Debey, Nilles, Lehmann, & Gay, 2012). Contamination may be more widespread and sustained than projected in models because uranium bucky-balls, or durable transportable structures created from the use of salt-water to cool melted fuel, may have increased transportability of radionuclides, potentially invalidating dispersion projections (Priyadarshi, Dominguez, & Thiemens, 2011).

Lack of ongoing sampling on land and in fresh and ocean water may lead scientists to underestimate the long-term effects of the disaster on the environment, particularly the ocean. The contamination from the Daiichi site has continued unrelenting for three years and could increase exponentially if any of the reactor buildings suffered complete loss of fuel containment. The "German Risk Study, Phase B" found that a core meltdown accident could result in complete failures of all structural containment, causing melted fuel to exit the reactor foundation within five days (cited in Bayer, Tromm, & Al-Omari 1989). Moreover, the study found that even in the event of an intact building foundation, passing groundwater would be in direct contact with fuel, causing leaching of fission products. Strontium leaches slower than cesium. A follow-up German study, "Dispersion of Radionuclides and Radiation Exposure after Leaching by Groundwater of a Solidified Core-Concrete Melt," predicted that strontium contamination levels would rise exponentially years after a full melt-through located adjacent to a river (Bayer, Tromm, & Al-Omari, 1989). The study's experimental conditions are roughly similar to Daiichi's site conditions, including groundwater emptying into an adjacent river, whereas Daiichi is physically situated above an underground river emptying into the sea. The study predicted concentrations of Strontium-90 in river water would spike relatively suddenly, but maintain extraordinarily high levels of contamination for years: "The highest radionuclide concentration of approx. 10^{10} Bq/m^3 is reached by Sr-90 after some 5000 days." It is noteworthy that TEPCO reported an exponential increase in the Strontium-90 level in ground water beginning in July 2013, with levels continuing to rise through early 2014. The accident at Daiichi may fit the German melt-through scenario. An exponential increase in the level of Strontium-90 contamination would invalidate models of projected ocean contamination relying on samples collected in 2011. For instance, a study on expected strontium contamination levels off Japan published in 2013 titled "90Sr and 89Sr in seawater off Japan as consequence of the Fukushima Daiichi nuclear accident" predicted consequences based on strontium-cesium ratios derived from 2011 samples (Casacuberta, Masqué, Garcia-Orellana, Garcia-Tenorio, & Buesseler, 2013).

The melt-throughs that occurred at Fukushima Daiichi may produce unprecedented contamination, which will bioaccumulate and biomagnify in the food chain without check in the absence of ongoing monitoring. Prior to Fukushima, the Pacific Ocean measured ½ to 2 Becquerels per liter, according to Ken Buesseler of Woods Hole Aquarium (Buesseler, 2013). In October of 2013, Buesseler predicted that the Pacific Ocean could measure up to 30 Becquerels per liter on average, although he was unclear about whether that figure applied exclusively to cesium. Other radionuclides are also of concern, including strontium, uranium, and plutonium, among others. Tritium, for example, is known to bioaccumulate in phytoplankton, which is a keystone species. Biomagnification of tritium in phytoplankton poses a persistent and toxic contaminant with intergenerational effects (Jaeschke & Bradshaw, 2013). The ongoing ocean contamination from Daiichi is threatening the global commons of the ocean.

Limited Liability for a Human Wrought Disaster

In 2013 *The Asahi Shimbun* projected that approximately 54,000 people would not be able to return home by 2017 because they formerly inhabited what is now designated as a five year exclusion zone (beginning March 2011) set up by the Japanese government for areas with 50 millisieverts a year or more of radiation exposure (Kansai, 2013). Areas measuring between 20 and 50 millisieverts a year were designated as areas with living restrictions. Areas measuring fewer than 20 millisieverts a year of annual exposure were designated as habitable zones and preparations were made for lifting evacuation orders in these areas.

Japan has essentially upped its national exposure level from one to 20 millisieverts a year, while allowing partial habitation in areas with up to 50 millisieverts. In comparison, the Soviets set the Chernobyl exclusion zone at five millisieverts a year ("Japan Groups Alarmed," 2011). Through the creation of these zones, the Japanese government is encouraging people to return to heavily contaminated areas. Months after the zones were delineated, Japan announced that it would be changing the way radiation levels were monitored in contaminated areas by moving to a badge system, away from atmospheric monitoring. According to a November 9, 2013 report from *The Asahi Shimbun*, the badges underestimated exposure levels by seven times when compared to the atmospheric monitoring technique:

> The Nuclear Regulation Authority has drafted a proposal to accelerate the return home of Fukushima nuclear disaster evacuees by using radiation readings that tend to be lower than the ones now officially used. The NRA wants residents to take radiation measurements with dosimeters instead of relying on the current government system of determining levels through aircraft monitoring.... ("Lower Radiation Readings," 2013)

This change essentially increases permissible exposure levels.

After the accident, the Japanese Diet legislated "Establishment of a Nuclear Damage Compensation Facilitation Corporation," which was instituted in September of 2011 to dispense compensation funds to the more than 71,000 people who were evacuated due to the nuclear disaster. However, compensation agreements for citizens forced to evacuate by the disaster were disclosed slowly, creating uncertainty and frustration among evacuees. It was not until late 2013 that the Japanese media outlined details of the compensation program, as illustrated here by an account published by *The Asahi Shimbun* in December of that year:

> A family of four from the difficult-to-return zones will be eligible to receive 106.75 million yen overall. . . . It is also estimated that a family of four from zones where residences are restricted--where radiation levels are between 20 and 50 millisieverts per year--will receive 71.97 million yen. A family of the same number from areas where preparations will be made to lift the current evacuation order, which cover regions with an annual radiation level of 20 millisieverts or lower, will be able to receive 56.81 million yen, according to the estimates. (Negishi, Ozawa, Seino, & Otsuki, 2013)

Presumably, disbursement will end all other compensation payments. Currently, policy sets a one year limit on compensation payments when evacuation orders are lifted, although promises have been made to extend the one year limit under special circumstances ("Panel Willing," 2013).

Many citizens are dissatisfied with the proposed compensation plan, arguing in particular that areas with atmospheric radiation levels up to 20 millisieverts a year are unsafe for human habitation. Former residents are concerned that they are being required by economic necessity to return to contaminated areas where local clean-up will be slow and ineffective and that new "badge" measurement devices for atmospheric levels will understate actual exposure through inhalation and ingestion of radionuclides. Worries about decontamination are particularly strong. Fraud and corruption have been pervasive among contractors used for decontamination and decontaminated areas have been quickly re-contaminated. Contractors have been caught dumping radioactive waste, calling into question the legitimacy of the entire decontamination effort, as illustrated by *The Asahi Shimbun's* series on "Crooked Clean-Up" (Kihara & Aoki, 2013).

Responsibilities for financing and executing decontamination have been highly contested because local communities are often held responsible for executing and paying for decontamination. In August 2011 Japan passed the "Act on Special Measures Concerning the Handling of Radioactive Pollution," which delineates responsibility for clean-up. The Ministry of the Environment is designated as responsible for categorizing contamination zones. Two zones of contamination were carved out by the act. First is the **"Special Decontamination Area,"** for which the Japanese government is responsible for clean-up (Ministry of the Environment, 2013). It includes 11 municipalities in the (formerly) restricted zone or planned evacuation zone less than 20 kilometers away from the NPS. It encompasses areas with cumulative doses in excess of 20 millisieverts a year. Second is the **"Intensive Contamination Survey Area,"** for which local governments are responsible for clean-up, although they are supposed to have access to financial and technical supports from the national government. The Intensive Contamination Survey area includes land from 104 municipalities where the dose rate is over 0.23 Sv/h (equivalent to over 1 mSv/y of additional dose), but less than 20 millisieverts a year.

The problem is that these designations designate local responsibility for decontamination in areas under 20 millisieverts a year when the international standard is one millisievert, and they fail to anticipate the long-term dispersion and bioaccumulation of radionuclides in the environment. Distribution of radionuclides is not uniformly homogeneous, nor is it static. Radionuclides in the water or attached to dust and pollen in the atmosphere circulate in plumes, depositing concentrations variably, depending upon geographic and atmospheric conditions. Radionuclides also circulate as they are absorbed by flora and inhaled and ingested by fauna. Bioaccumulation and biomagnification amplify concentrations in animals high on the food chain, such as humans. Decontamination is a never-ending process that has yet to be successfully mastered, as clean-up at Hanford in the US illustrates. Communities simply do not have the resources to de-contaminate properly, resulting in storage of radioactive waste (including dirt and leaves) in bags in close proximity to living spaces. The Japanese government's policy responses to clean-up and evacuation have essentially transferred many of the costs and risks of decontamination areas to local officials and citizens.

Some residents have resisted, particularly when prompted by concerns for their children's safety. Children may have particular biological and psychological vulnerabilities to nuclear disasters. By February of 2014, there were 75 confirmed or suspected thyroid cancer cases among 270,000 Fukushima Prefecture individuals screened, who were 18 or under at the time of the disaster (Nose & Oiwa, 2014). The screening committee claimed the Fukushima disaster was an unlikely cause ("Eight More," 2014). However, the observed frequency of thyroid cancer and nodules exceeds established incident rates. For example, the prevalence of thyroid nodules in children typically ranges from 0.2-5.0 percent (Gerber & Meyers, 2013), while in Fukushima, 42 percent of 133,000 children were found to have thyroid nodules and cysts (Haworth, 2013). A study measuring thyroid exposure to Iodine-131 conducted between April 12, 2011 and April 16, 2011 and published in *Research Reports* found "extensive measurements of the exposure to I-131 revealing I-131 activity in the thyroid of 46 out of the 62 residents and evacuees measured" (Tokonammi, Hosoda, Akiba, Sorimachi, Kashiwakura, & Balonov, 2012).

Children from Fukushima are exhibiting worrying symptoms. For example, a 2013 survey of 178 children enrolled in nursery schools at the time of the disaster in Northeast Japan found that many suffer from problems including "dizziness, nausea, headaches, swearing and reticence, among other symptoms" ("1 in 4 Disaster," 2014). The survey, which employed the internationally recognized Child Behavior Checklist (CBCL), suggested that "25.9 percent of children in Iwate, Miyagi and Fukushima Prefectures were diagnosed as being in need of medical care" because of trauma from the disaster. It is unclear whether the reported dizziness, nausea and headaches were from stress and anxiety, or were symptoms of exposure to chronic low-level radiation.

Parents concerned about their children's safety have pressed for evacuation support for children inhabiting zones deemed officially safe and habitable, with levels of exposure from one to twenty millisieverts a year. For example, parents and global anti-nuclear activists joined together in the Fukushima Collective Evacuation Trial Team to sue for evacuation of children from the city of Koriyama, which has been deemed safe for habituation. On April 24, 2013 the Sendai High Court handed down its ruling that the city of Koriyama had no legal responsibility to evacuate children at compulsory elementary schools and junior-high schools despite court-acknowledged radiation levels in the city exceeding levels deemed safe prior to the disaster. Unfortunately for the plaintiffs, the court ultimately denied government responsibility for evacuations, shifting responsibility to parents (The Associated Press, 2013). The Fukushima Collective Evacuation Trial Team contested the judge's decision, arguing that it shifted all responsibility for risk to the victims of the disaster (The Fukushima Collective Evacuation Trial Team, 2013). It appears from the ruling of the Sendai high court that cities have no legal responsibility for removing residents at risk from contamination, although cities are responsible for clean-up of inhabited Intensive Contamination Survey Area[s] under the "Act on Special Measures Concerning the Handling of Radioactive Pollution."

Japanese citizens have also sued manufacturers of Fukushima Daiichi reactors for losses. 1,400 plaintiffs are suing General Electric Japan, Hitachi Ltd. and Toshiba Corp. for the emotional distress they suffered from the reactor meltdowns:

> The complainants insist that the nuclear damage compensation facilitation law is unconstitutional, and therefore invalid, because it prevents people from filing suits that question the responsibility of reactor makers.... "It is a lawsuit designed to drag the makers of the reactors out of hiding," said Hiroyuki Kawai, who is also representing TEPCO shareholders in a lawsuit they have filed against the utility's executives. The amount of claimed compensation was set low--just 100 yen ($1) per person--because the lawsuit is primarily intended to force the reactor makers to accept their responsibility. ("1,400 Sue," 2014)

However, this suit may also be lost because Japanese citizens' avenues for compensation are limited by the terms of the nation's nuclear liability legislation, **The Act on Compensation for Nuclear Damage**, enacted in 1961. The act limits liability to the plant operator, freeing manufactures from liability (Osaka, 2012). However, it does leave open the scope of damages available for compensation from the operator so long as the nuclear event was not caused by war, sabotage, or a grave natural disaster. This act doesn't specify a statute of limitations, therefore the Japanese Civil Code applies, which ordinarily limits tort claims to twenty years, although exceptions have been made when injuries are accumulative and latent, allowing victims to file within three years of development of the injury (Osaka, 2012). In January 2013, the Japanese media reported that TEPCO and the Nuclear Damage Liability Facilitation Fund were limiting damage claims to a three year period beginning when the claimants receive application documents for compensation (Request for Compensation, 2013).

In principle, the Japanese Diet's designation of Fukushima as a "man-made" disaster exposed TEPCO, operator of the Daiichi site, to potentially unlimited financial liability for damages, including emotional distress and property losses. However, TEPCO's private insurance at the time of the disaster was limited to JPY 120 billion with the Japan Atomic Energy Insurance Pool, a pool for nuclear operators that is reinsured with other atomic energy insurance pools around the world (Vasquez-Maignan, 2011). After the disaster TEPCO was ineligible for insuring further losses through the pool ("Nuclear Power Plant Insurance," 2012). TEPCO became fully reliant on its protection through Japan's national **Act on Indemnity Agreements for Compensation of Nuclear Damage**, referred to as the "Indemnity Agreements Act," which was also passed in 1961. The Indemnity Agreement authorized the Japanese government to offer nuclear operators indemnity agreements covering uninsured losses. Under this act, the Japanese government indemnifies losses suffered by TEPCO and other nuclear operators for nuclear damage compensation caused by a visiting ship, earthquake, volcanic eruption, normal operations, and "other" conditions.

In practice, the Act on Compensation and the Act on Indemnity Agreements for Compensation of Nuclear Damage shift losses from TEPCO to the government and people of Japan. First, costs associated with damages not covered by private insurance (e.g., from earthquakes and tsunamis) are indemnified by the government. Second, costs exceeding the private insurers' JPY 120 billion limit are also indemnified by the Japanese government. Through these indemnity contracts the Japanese government ultimately assumes the liability risks for uncovered costs. Financial losses are ultimately transferred to citizens through budgetary compensation measures, such as tax hikes, and through special designation zones that assign clean-up to local communities. Statutes of limitations limit opportunity for filing for nuclear damage, despite the fact that radiation damage produces trans-generational effects, which will be discussed presently. Individuals therefore shoulder direct as well as indirect costs from nuclear damage.

Citizens of Japan shouldered Fukushima disaster costs as the government assumed TEPCO's liabilities when TEPCO faced bankruptcy in late 2011 (Uranaka, 2012). In July of 2012, the Japanese government made a 1 trillion yen direct equity investment into TEPCO using tax payer funds, affording the government a voting right in excess of 50 percent. The stocks purchased through this equity investment are owned publicly by the government Nuclear Damage Liability Facilitation Fund. Additionally, the government announced it would provide 2.5 trillion yen in compensation to victims rather than forcing TEPCO to go into bankruptcy and liquidate assets. TEPCO was allowed to increase electricity rates by 8.46 percent for household consumers and 14.9 percent for corporations, which effectively transfers the direct financial cost for paying for the disaster to rate holders, as well as citizens (Ohira & Fujisaki, 2012).

According to an October 27, 2013 article in *The Asahi Shimbun*, the Japanese government policies tacitly accept TEPCO's refusal to pay clean-up costs:

> **"Documents show government tacitly accepts TEPCO's refusal to pay for cleanup"** Tokyo Electric Power Co., operator of the embattled Fukushima No. 1 nuclear power plant, declared early this year that it will not repay radioactive cleanup costs in Fukushima Prefecture, forcing taxpayers to shoulder the burden, The Asahi Shimbun has learned. The government, which did not release TEPCO's statement, apparently accepts the refusal, in a tacit understanding to prevent the cash-strapped utility from being driven into bankruptcy. Documents obtained by The Asahi Shimbun through a freedom of information request showed that TEPCO in February made clear its intention not to pay the full cleanup costs. ("Documents Show," 2013)

TEPCO's costs for stabilizing the reactors were estimated 5.8 trillion yen in late 2013 (Hasegawa, 2013). The Japanese government announced in December of 2013 it would use future sales of TEPCO shares held by the Nuclear Damage Liability Facilitation Fund to pay for clean-up costs (Fujisaki & Ebuchi, 2013). The shares are expected to be sold in the late 2020s or 2030s "after the utility turns around its finances." This decision indirectly commits the Japanese government to maintaining TEPCO's share value. The Japanese government also increased lending to TEPCO up to 10 trillion yen and dedicated 350 billion yen in the fiscal 2014 budget to help fund decontamination work costs around the Daiichi site (Nogami & Fujisaki, 2013).

As these examples illustrate, the full costs of the disaster are being absorbed by the national and local governments in Japan. Japanese citizens will ultimately pay for the disaster through local decontamination costs, higher taxes and electricity rates, reduced property values, and through health injuries sustained from radiation exposure across time. Limited liability shifts risk to those who bear externalities. A short statute of limitations further accentuates risk shift. Risk shifts still more to citizens by aggressive efforts to re-establish nuclear operations in Japan. In sum, limited liability is fundamentally market distorting because wrong-doers are allowed to externalize their risks when not required to make full restitution. In the nuclear industry, national legislation and international conventions restrict liability from accidents while permissible dose standards limit liability from routine emissions.

Externality: Liberal Personhood

Private property rights, including ownership and governance of one's own body, are casualties of nuclear power. Individual bodily security and personal assets alike are threatened by nuclear contamination. Yet, as explained above, limited liability provisions absolve those who design and operate nuclear power plants from absorbing the full costs of contamination from routine operations, waste storage, and nuclear accidents. Therefore the health effects of contamination and the financial costs of clean-up are typically absorbed by the public, either directly in the form of personal injuries or indirectly in the form of higher taxes necessitated by government bailouts and clean-up. This section addresses the risks and costs of nuclear externalities by examining how living in a contaminated zone destroys livelihoods.

People living in a radiation contaminated zone confront devalued property and face risks to their health and mental well-being. Radioactive elements produced and/or released by nuclear fission include radioactive isotopes of cesium, iodine, strontium, as well as all the transuranic elements, such as plutonium. Through wind and water, radionuclides released by containment failures can contaminate areas thousands of miles from fission events. Fallout of radioactive elements bioaccumulates in plants and animals. Biomagnification occurs over time as predators accumulate the radio-isotopes of their prey. Contaminated animals suffer compromised immune systems, accelerated aging, and are more likely to die of a range of diseases. Moreover, genetic and epigenetic damage wrought by decaying radionuclides are transmitted across generations. Understanding these processes is vital because with understanding comes the realization that international radiation protection guidelines are inadequate for protecting human health.

Nuclear fallout emits ionizing radiation - including gamma, beta, alpha, and neutron radiation - capable of stripping electrons from the atoms that compose DNA. Fallout of radioactive particles, such as cesium and uranium, can be inhaled or ingested through contaminated food and water. Some radionuclides – such as tritium – can penetrate skin. Radionuclides that are inhaled, ingested, or penetrate skin pose higher exposure risks because many among them are stored within the body (i.e., "bioaccumulate"). Some radionuclides are especially likely to bioaccumulate because living organisms – including both plants and animals – confuse them with elements needed for vital processes. So, for example, both plants and animals misrecognize radioactive isotopes of cesium in soil (e.g., Cesium-137) as non-radioactive potassium. Radio-strontium is likewise confused with calcium. Strontium can be stored in the brain and then mobilized to trigger neurotransmitter release, replacing calcium in neurochemical processes (Xu-Friedman & Regehr, 1999). Bioaccumulation of chemically toxic and radioactive elements in human bodies poses significant long-term effects to human health and reproduction.

Bioaccumulation is always worst for animals at the pinnacle of their food chain because of biomagnification. Biomagnification occurs as predator animals consume the radionuclides bioaccumulated in their prey. Risks associated with biomagnification have been understood for decades. In 1962, Harold Knapp described how radioiodine from a single deposition in pasture-land bioaccumulates and biomagnifies, producing substantial and injurious radiation doses for children consuming milk (Kirsch, 2004). Radioiodine bioaccumulation in the thyroid gland can disrupt normal development, in addition to causing cancer.

The disease processes caused by radiation bioaccumulation are complex because radiation both causes disease, and increases susceptibility to them, by suppressing the immune system and accelerating aging processes. Ingested radionuclides, such as uranium and strontium, are chemically toxic while radioactive decay processes of beta, gamma, and alpha radiation produce micro level biological damage. For example, beta particles - (shed by ingested radionuclides such as strontium and cesium) – are accelerated electrons/positrons that disrupt chemical bonds as they pass through cells, promoting creation of unstable and chemically reactive atoms, called radicals. Damaged DNA and the increase in free radicals result in genomic instability, even among cells escaping direct exposure to radioactive decay. Genomic instability leads to increased error in cell production. Error compromises the body's biological repair mechanisms and immunological defences. Inflammation results from these insults, rendering the individual organism susceptible to other disease processes. Cancer is but a single disease among many caused by chronic exposure to radionuclides.

Research suggests that ionizing radiation adversely alters the highly radiosensitive T-cell system, which is vital to the immune response, and radiation damages mitochondrial DNA, which is particularly susceptible to mutation from ionizing radiation (Lutz-Bonengel, Brinkmann, Forster, Forster & Willkomm, 2002). Japanese research on the immunological health of atomic bomb survivors concluded that "radiation on T-cell immunity resemble effects of aging on the immune system" (Kusunoki et al., 2010). Radiation ages the immune system, accelerating entropy in bodily processes. Radiation also compromises reproductive health through the transgenerational transmission of mutations. For example, a 2002 study found that excess radiation exposure accelerates point mutations in mitochondrial DNA that are transmitted across generations, even among people accustomed to higher than average background radiation (Lutz-Bonengel, Brinkmann, Forster, Forster, & Willkomm, 2002). Laboratory research on mice exposed to chronic radiation documented intergenerational effects from exposure, including "increased instability of repeat-DNA sequences" in descendants of affected mice, due in part to increased "mutational mosaicism" of the germ line (Dubrova, Plumb, Guiterrez, Bolton, & Jeffreys, 2000, p. 37). A growing body of research exploring environmental genomics substantiates geneticists' warning in the 1956 BEAR report that any additional exposure to radiation increases genetic mutations, which are heritable in germ-line cells. Early scientists studying radiation, such as H. J. Mueller and J. W. Gofman warned then that human health and reproduction are at risk from the transgenerational accumulation of mutations caused by exposure to nuclear fallout from atmospheric testing (Nadesan, 2013).

The Soviets encapsulated ionizing radiation's complex deconstructive processes in the disease syndrome known as Chronic Radiation Syndrome. This syndrome destroys health and well-being, as documented by Kate Brown's *Plutopia: Nuclear Families, Atomic Cities, and the Great Soviet and American Plutonium Disasters*, which describes the affliction of chronic radiation syndrome in the people of a village of Muslumovo, located in the southern Russian Ural Mountains. The village is located downstream of the Maiak plutonium plant, which manufactured plutonium-based bomb cores beginning in 1948. The plant used the Techa River to dispose of high level radioactive waste. The numerous villages along the river were not informed, despite using its water for drinking, cooking, and bathing. Some of the villages along the contaminated river were eventually evacuated, but not Muslumovo. In the early 1990s, a paediatric doctor working in the village, named Glufarida Galimova, was puzzled by and began investigating the high incident of strange disorders, including hydrocephalic children, children with cerebral palsy, missing kidneys, extra fingers, anaemia, fatigue, and weak immune system. Galimova found that more than half of the children suffered pathologies. By 1999, 95 percent of children born in the village suffered from genetic disorders while 90 percent of the entire childhood population experienced chronic anaemia, fatigue and/or immune disorders. After examining medical records, Galimova concluded that only seven percent of adults in the city could be considered healthy.

Galimova was not the only doctor observing irregularities. By 1962, the Soviet Institute of Bio-Physics, called FIB-4, had begun systematic study of the health of the Muslumovo population. When Galimova inquired about the cause of the birth defects documented by FIB-4 she was told the culprit was alcohol, although the FIB-4 group had diagnosed 935 people on the Techa River with Chronic Radiation Syndrome, a new disorder labelled in 1950 by a doctor at the plutonium plant. Early symptoms of the disorder included headaches, sharp pains in bones and joints, and chronic fatigue. Blood changes, including severe anaemia, typically followed these symptoms. Longer term symptoms included heart disease and markedly slowed gait.

Alexey V. Nesterenko, Vassily B. Nesterenko, and Alexey V. Yablokov (2009) observe in their Introduction to *Chernobyl: Consequences of the Catastrophe for People and the Environment* that former UN Secretary-General Kofi Annan calculated that at least 7,000,000 people were adversely impacted by that disaster. Their review of 5,000 medical and scientific studies concluded there were 985,000 deaths from Chernobyl between 1986 and 2004, primarily from cancer, heart and other circulatory diseases, and excess infant mortality. Perhaps most troubling of all, they argue only 20 percent of children living in the Chernobyl contaminated areas of Belarus, Ukraine, and European Russia are considered healthy. Research from the region suggests Chernobyl radiation bioaccumulated in children's bodies and affected their genomic stability. Yuri Bandazhevsky found that children contaminated with Cesium-137 producing 50 disintegrations per second (Becquerels) per kilogram of body weight suffered irreversible heart damage (Starr, 2012). Anna Aghajanyan and Igor Suskov (2009) found that male Chernobyl liquidators and their children had increased aberrant genome frequencies, suggesting transgenerational genomic instability as a consequence of radiation exposure. A 2008 review of findings on genomic damage in children published in *Mutation Research* concluded that Chernobyl-radiation exposed children suffered consistently increased chromosome aberration and micronuclei frequency (Fucic et al., 2008). Another study found that that chronic low-dose exposure to radiation from Chernobyl caused increased rates of neural tube-defects and conjoined twins (Wertelecki, 2010).

Biological effects on health and reproduction from Chernobyl fallout occurred widely. Almond, Edlund and Palme (2009) found cognitive effects, particularly retardation among Swedish children exposed in utero to Chernobyl fallout at 8 to 25 weeks of gestation. The critical period for neurogenesis roughly corresponds to this time period. Another study documented post-Chernobyl mortality increases in infants in Germany (Körblein & Küchenhoff, 1997) and also in the elderly and auto-immune compromised in the U.S. (Gould & Goldman, 1991).

Children are especially vulnerable because they have inherited both parents environmentally acquired germ line cell damage and their cells are dividing quickly. Embryos and fetuses exposed to radiation are even more vulnerable. Nowakowski and Hayes (2008) explore the myriad effects of radiation on early brain development (i.e., neurogenesis), which include double-strand breaks of DNA impacting cell proliferation and migration during critical periods of early brain development. They conclude that early fetal development is particularly susceptible to effects of relatively low levels of exposure to radionuclides from nuclear accidents, among other sources of exposure.

People living in radiation contaminated zones risk premature aging, compromised immune responses, cancer, and declining reproductive health. The epidemiological findings have been substantiated by research on health effects from radiation exposure. For example, findings on premature aging and genetic disorders among people in contaminated zones are consistent with a general pattern of premature aging found in adult survivors of childhood cancer. A 2013 study published in *The Journal of Clinical Oncology* reported that survivors of cancer treated by radiation and chemotherapy experienced significant premature aging, manifested in symptoms such as frailty, slowed gait, low muscle mass, and weakness (Ness, Krull, Jones, Mulrooney, Armstrong, Green, & Hudson, 2013). Radiation therapy, chemotherapy and the cancer disease process are all implicated in causing the range of symptoms, although respective influences have yet to be differentiated.[IV] Other studies find that high LET radiation – which includes neutrons and alpha particles – causes more radical formation within a cell than low LET radiation, causing premature aging. Research on mice reveals that that additional exposure to high LET radiation produces mitochondrial dysregulation, prolonged oxidative stress and premature aging, and premature senescence (see Datta et al., 2012; Suman, et al., 2013; Trani et al., 2010). Mitochondrial DNA susceptibility to radiation-induced mutation is of particular concern because mitochondrial DNA is inherited only from the mother, increasing genomic vulnerability.

In sum, the studies described here have documented empirically accelerated aging and disease processes in people and animals exposed to increased radiation levels. Accelerated aging and disease processes derive from increased exposure to ionizing radiation. Cancer and genetic disorders are the most obvious symptoms of chronic radiation exposure, but the actual range of effects is multifaceted because excess radiation exposure fundamentally deconstructs the body by causing DNA damage and impinging against repair processes. People who live in radiation contaminated zones are essentially living in environments that deconstruct their genome, their cell integrity, and the genetic health of their offspring. Nuclear power threatens personal property rights and the pursuit of health and happiness. It is also antithetical with democracy itself.

Externality: Transparency and Democracy

In the months after the disaster, Yukio Edano, Chief Cabinet Secretary of Japan, acknowledged the importance of risk communication and pledged to transform his nation into a "risk-resistant society": "We will make efforts to improve transparency and more readily share information," Edano declared to the World Economic Forum's Risk Response Network meeting (cited in Tonkin, 2011). Transparency is widely considered foundational to good political and economic governance because it promotes responsibility to uphold laws of a democratic society and facilitates accountability (Florini, 2007). High transparency is believed to overcome informational asymmetries that enable insiders to make self-interested decisions that might adversely impact "outside" stakeholders. Transparency is the mechanism for promoting informed judgment in the market as well. This section turns to examine deliberate efforts at censorship in Japan that threaten transparency and democratic governance.

As previously explored in this chapter, the national Diet of Japan's report on Fukushima documented that the lack of transparency in crisis management directly compromised citizen safety. The Japanese news media have also chastised the government for lack of transparency. For example, in December of 2013, *The Asahi Shimbun* ran an editorial charging that Japan's bureaucratic secrecy hindered the disaster response, using the example of B.5.b. (Okuyama & Sunaoshi, 2013). B.5.b is a highly classified U.S.-developed contingency plan for a catastrophic nuclear plant event given to the Japanese Nuclear and Industrial Safety Agency (NISA) prior to the March 2011 disaster. It remained highly classified during the March 2011 Daiichi explosions as only a handful of officials were privy to information that may have improved the insufficient disaster response (Okuyama & Sunaoshi, 2013). This is not the first time secrecy has been cited as hindering the disaster response. High ranking government officials failed during the disaster to disclose publicly their information about the characteristics and directionality of radioactive plumes, which had been modeled by the SPEEDI system (i.e., the System for Prediction of Environmental Emergency Dose Information, Tabuchi, Bradsher, & Pollack, 2011). In July of 2011, the Atomic Energy Society of Japan publicly criticized the Japanese government and TEPCO for delays in reporting SPEEDI data to the public, arguing that the lack of transparency may possibly have increased radiation exposure risks ("Nuclear Accident Disclosure," 2011). The Japanese Diet report reached similar conclusions. Secrecy was long-rationalized by the LDP government and Japan's utilities as a strategy for reducing resistance to nuclear energy ("TEPCO Seeks More Government Support," 2012).

Japan's record on political transparency has substantially deteriorated since Edano made his pledge in 2011. In 2013 Japan's LDP party pushed through a highly controversial state secrets bill. The bill has been vocally opposed by political opposition to the LDP, by editorials in the Japanese media, and in excoriating comments made by well-known Japanese citizens. The law stipulates harsh penalties for whistleblowers and fails to require government disclosure of what becomes secret (Yamaguchi, 2013). Critics charge it could undermine Japan's democracy.

Many in Japan see the secrecy bill as a direct threat to Japan's status as a peaceful democratic nation. One survey found more than 80 percent of Japanese citizens distrust the law, feeling it will be used by government to hide corruption and troubling information (Adelstein, 2013). Discontent with the bill is high among Japanese officials of opposition parties and within professional organizations. "Japan already has a very weak freedom of information act which this will cripple," said Yutaka Saito, a member of the Japan In-House Lawyers Association task force. "The bill takes everything bad about national security laws in the U.S. and then removes all the safeguards and checks" (Adelstein, 2013).

Well-known Japanese citizens have also raised alarm about the bill's potential effects. Nobel laureates, Toshihide Masukawa and Hideki Shirakawa, spearheaded a public letter of protest signed by 3,000 academics, declaring support for "the pacifist principles and fundamental human rights established by the constitution" while calling simultaneously for the law's immediate rejection" (Nader, 2014). Non-fiction writer Kunio Yanagida talked to Hiroshi Dai, Senior Member of the "Open Newspaper" Committee about his concerns with the bill given Japan's historically "secretive" bureaucratic culture:

> In the Japanese bureaucracy, it is common to hide not only information concerning national defense or foreign affairs but

to keep general information about any matter a secret. Since I was involved in the government's Investigation Committee on the Accident at the Fukushima Nuclear Power Stations and the Ministry of the Environment's panel for the Minamata disease issue, I strongly feel the negative effects of the secretive culture. . . . Bureaucrats hide their inconvenient truths while making up things that never existed. The bureaucracy today still functions on the principle that things are fine as long as they add up in the paperwork. This thinking once played a major part in the former Japanese Imperial Army. The number of special secrets will increase just like military secrets did before Japan went into World War II.... ("Writer Calls for Change," 2014)

Yanagida went on to warn that the bill could actually criminalize free speech. He observed that LDP's Shigeru Ishiba's comment comparing "protest rallies" to "terrorism" illustrates growing intolerance for free speech among government officials. He also warned that journalists could be punished for reporting on state secrets.

International and national media organizations have echoed these concerns. Reporters without Borders condemned the bill on November 27, 2013, for violating a free press (Adelstein, 2013). Within Japan, the mainstream media have demonstrated their concern about the new law in prominent editorials decrying its effects on democracy and civil society, as exemplified by this December 2013 editorial in *The Japan Times* titled "Government without Oversight":

People have the right to know what their government is doing. Ensuring this right is the foundation of democracy. The state secrets bill, which the Abe administration Thursday rammed through the Upper House Special Committee on National Security for enactment, undermines this foundation because it blocks citizens" access to an extremely large amount of government-held information. This also means that lawmakers" access to important information held by bureaucracy will be blocked. Citizens and lawmakers should be aware that the bill will greatly change the nature of Japanese politics because it will severely limit the powers of people's representatives and the Diet itself despite the fact that Article 41 of the Constitution says, "The Diet shall be the highest organ of state power, and shall be the sole lawmaking organ of the State." Japan's democracy is now in a deep crisis. ("Government without Oversight," 2013)

A more trenchant editorial published also in *The Japan Times* in 2013 described the bill as transforming the nation into the "new Uzbekistan of press freedom in Asia":

The first rule of the pending state secrets bill is that a secret is a secret. The second rule is that anyone who leaks a secret and/or a reporter who makes it public via a published report or broadcast can face up to 10 years in prison. The third rule is that there are no rules as to which government agencies can declare information to be a state secret and no checks on them to determine that they don't abuse the privilege; even defunct agencies can rule their information to be secret. The fourth rule is that anything pertaining to nuclear energy is a state secret, which means there will no longer be any problems with nuclear power in this country because we won't know anything about it. And what we don't know can't hurt us. The right to know has now officially been superseded by the right of the government to make sure you don't know what they don't want you to know..... The law has been compared to the pre-World War II Peace Preservation Law, which was used to arrest and jail any individual who opposed the government party line. (Adelstein, 2013)

As illustrated by these passages, many Japanese media outlets took a strong stance against the law, fearing loss of democratic governance and the criminalization of free speech. The Japan Federation of Bar Associations publicly raised concerns that the new law would increase the government's tendency to censor nuclear information (Okuyama & Sunaoshi, 2013). One representative from that organization suggested the law could be used to censor radiation readings. Other observers note that the law could compromise nuclear safety, as illustrated by this article in *The Asahi Shimbun*: "State Secrets Law Raises Concern about Safety of Nuclear Power Plants":

There is growing concern that the government may be tempted to keep sensitive information on the safety of nuclear power plants under wraps once the state secrets protection law goes into force.... One reason for this is that the legislation leaves unclear what matters will be designated as state secrets and who will have authorization to determine what should be withheld from the public. "A tendency to hold back on vital information left nuclear power plants vulnerable to earthquakes and tsunami, resulting in the nuclear disaster," said Yutaka Saito, a member of the federation's task force on problems related to information. "We cannot fully engage in discussion about safety if information is withheld." ("State Secret Law Raises Concern," 2013).

As illustrated by these examples, Japan's new state secrets law has widely been challenged for posing direct risks to government accountability and transparency.

Among the most disturbing instance of failed transparency is the purported "secrecy agreement" established between the IAEA and Fukushima Prefectural

University of Medicine on health findings from radiation exposure. Accordingly, *The Tokyo Shimbun* reported "Secret Designation Clause Fukushima IAEA, Fukui Share Private Information," explaining that "private" data shared by the IAEA and Fukushima Prefectural University Hospital about radiation health effects will not be reported publicly ("Secret Designation Clause," 2013). Concern that radiation health effects will be censored is historically well grounded. Shuntaro Hida, a Japanese doctor who survived the atomic bomb, has written and lectured widely about official and self-censorship of radiation effects in the wake of the Hiroshima bombing at the close of World War II. He explains that stigmatization from radiation exposure was so great that citizens would deny symptoms even when seriously impacted. Hida writes in his book, "Naibu Hibaku no Kyoi" (*The Threat of Internal Radiation Exposure*). "This is inherited by the second (children's) generation, and the third generation . . ." (cited in "Bomb Survivor Doctor," 2012). This same pattern of censorship can be found in the US as well (e.g., see Goliszek, 2003), as illustrated historically by the CIA's destruction in 1973 of records of secretive studies involving plutonium injections on human subjects, without their consent (Cockburn, 1994), and presently by the Nuclear Regulatory Commission's crisis communications campaign aimed at reducing all public concern about Fukushima fallout in the U.S. (Dedman, 2014).

The state secrets law legally institutionalizes censorship and reinforces its indiscriminate practice. In early January 2014, Prof. Toru Nakakita of Tokyo University quit administering the "Business Outlook" section of a NHK weekday broadcast, titled, "Radio Asa Ichiban" after being asked to avoid broadcasting a planned 30 minute segment examining the enormous financial damages from nuclear plant accidents ("NHK Radio Regular," 2014). The network was concerned about the influence the story might have on the ongoing Tokyo gubernatorial election. Nakakita explained his decision to the press: "We should have substantial debates precisely because it is the campaign period . . . . NHK reacted with excessive voluntary restraint, which shows a lack of awareness of the issue."

Noted American consumer activist Ralph Nader warned that growing secrecy and legislative control in Japan were bad omens for democratic rights:

> Draconian secrecy in government and fast-tracking bills through legislative bodies are bad omens for freedom of the Japanese press and freedom to dissent by the Japanese people. Freedom of information and robust debate (the latter cut off sharply by Japan's parliament in December 5, 2013) are the currencies of democracy. (Nader, 2014)

Nader concludes by noting that the "lessons of history beckon" in response to growing authoritarian trends in Japan and chronic secrecy in the U.S.

Conclusions

In March 2014, prosecutors dropped charges over the Fukushima nuclear disaster (Ryall, 2014). "Absolutely no-one is taking responsibility for this huge accident and when all these people are suffering," Aileen Mioko Smith, of Kyoto-based Green Action Japan, explained to *The Telegraph*:

> "The investigation clearly stated this was an accident created by humans, not a natural disaster, but the judicial system here has now decided to side with the powers-that-be," she said. "The government will be happy with the decision, but it is completely irresponsible," she said. "And I fear that failing to prosecute in this case will lead to another disaster in the future."

As argued so succinctly by Charles Perrow (2013), denial has been the conjoined twin of nuclear since the inception of the atomic bomb. Nuclear power produces myriad externalities and is fundamentally incongruous with liberal democracy. Nuclear power is market distorting because neither nuclear operators nor manufacturers are made responsible for the full costs of nuclear contamination from routine and accidental releases of ionizing radiation. Citizens assume the costs through higher taxes after accidents, through de-evaluation of property, and from health impacts, which can be transgenerational. People living in radiation contaminated zones are likely to incur more genetic damage, particularly to mitochondrial DNA. The free radicals unleashed by radioactive decay in the body accelerate the aging process. Ingested radionuclides are particularly likely to cause mutations and alter epigenetic operations, the effects of which can potentially give rise to future mutations in cell production. The Soviets studied the effects of "chronic radiation syndrome" in the village named Muslumovo and the Americans studied the effects of ingested radionuclides on children and prisoners. There is no safe level of additional exposure to radionuclides. Yet, this replicable finding is politically charged when examined closely and calls into questions formal assurances of "no excess risk." Indeed, the nuclear project in its entirety is inconsistent with the transparency required for healthy democracy. The LDP party has sacrificed free speech itself at the altar of nuclear energy.

References

1 in 4 disaster-hit children needs mental care for problem behavior: Study. (2014, January 27). *TheMainichi.*Available http://mainichi.jp/english/english/newsselect/news/20140127p2a00m0na013000c.html

1,400 sue makers of Fukushima nuke reactors. (2014, January 31). *The Asahi Shimbun.* Available http://ajw.asahi.com/article/0311disaster/fukushima/AJ201401310200

120 billion yen deposit insurance approval primary problem TEPCO. (2013, January 13). *Tokyo Shimbun.* Available http://www.tokyo-np.co.jp/article/feature/nucerror/list/CK2012011302100010.html

Adelstein, J. (2013, November 30). Japan: The new Uzbekistan of press freedom in Asia. *The Japan Times.* Available http://www.japantimes.co.jp/news/2013/11/30/national/japan-the-new-uzbekistan-of-press-freedom-in-asia/#.U2cUIV7K3yi

Aghajanyan, A., & Suskov, I. (2009). Transgenerational genomic instability in children of irradiated parents as a result of the Chernobyl nuclear accident. *Mutation Research, 671*(1-2), 52-57.

Almond, D., Edlund, L., & Palme, M. (2009). Chernobyl's subclinical legacy: Prenatal exposure to radioactive fallout and school outcomes in Sweden. Available http://people.su.se/~palme/QJErevisionJan23_09.pdf

Bayer, A., Al-Omari, I., & Tromm, W. (1989). *Dispersion of radionuclides and radiation exposure after leaching by groundwater of a solidified core-concrete* (No. KFK-4512). Available http://www.irpa.net/irpa8/cdrom/VOL.1/M1_97.PDF

Bomb survivor doctor continues to speak up about significance of internal exposure. (2012, January 23). *The Mainichi.* Available http://mdn.mainichi.jp/mdnnews/news/20120123p2a00m0na013000c.html

Bradsher, K., Tabuchi, H., & Pollack, A. (2011, April 12). Japanese officials on defensive as nuclear alert level rises. *The New York Times.* Available http://www.nytimes.com/2011/04/13/world/asia/13japan.html?_r=0

Buesseler, K. (2013, October 24). Japan's continuing nuclear nightmare: Experts discuss Fukushima and its aftereffects. Massachusetts Institute of Technology Center for International Studies. Available http://techtv.mit.edu/collections/mit-cis/videos/26614-japan-s-continuing-nuclear-nightmare

Carney, M. (2014, March 11). Fukushima disaster: Plan to send residents home three years after nuclear accident labelled "irresponsible." *Australian Broadcasting Corporation.* Available http://www.abc.net.au/news/2014-03-10/plan-to-send-residents-back-to-fukushima-meets-opposition/5311046

Casacuberta, N., Masqué, P., Garcia-Orellana, J., Garcia-Tenorio, R., & Buesseler, K. O. (2013). 90Sr and 89Sr in seawater off japan as a consequence of the Fukushima Dai-ichi nuclear accident. *Biogeosciences Discuss, 10*, 2039-2067.

Christoudias, T., & Lelieveld, J. (2012). Modelling the global atmospheric transport and deposition of radionuclides from the Fukushima Daiichi nuclear accident. *Atmospheric Chemistry & Physics Discussions, 12*, 24531-24555.

Cockburn, P. (1994, January 5). CIA "destroyed files on radiation victims": The public may never know full details of secret experiments on Americans during the cold war. *The Independent UK.* Available http://www.independent.co.uk/news/world/cia-destroyed-files-on-radiation-

victims-the-public-may-never-know-full-details-of-secret-experiments-on-
americans-during-the-cold-war-1397987.html

Datta, K., Suman, S., Kallakury, B. V., & Fornace Jr, A. J. (2012). Exposure to
heavy ion radiation induces persistent oxidative stress in mouse intestine.
PLOS One, 7(8), e42224.

De Alou, A. (2013, October 29) (Director). *Welcome to Fukushima* [Video/DVD].
Available https://www.youtube.com/watch?v=m7x3pdhn2ec

Dedman, B. (2014, March 10). US nuclear agency hid concerns, hailed safety
record as fukushima melted. *NBC News.* Available
HTTP://WWW.NBCNEWS.COM/STORYLINE/FUKUSHIMA-
ANNIVERSARY/U-S-NUCLEAR-AGENCY-HID-CONCERNS-HAILED-
SAFETY-RECORD-FUKUSHIMA-N48561

Documents show government tacitly accepts TEPCO's refusal to pay for cleanup.
(2013, October 27). *The Asahi Shimbun.* Available
http://ajw.asahi.com/article/0311disaster/fukushima/AJ201310270048

Dubrova, Y. M., Plumb, B. Gutierrez, E. Boulton, and A. Jeffreys (2000, May 4).
Genome Stability: Transgenerational Mutation by Radiation, *Nature,* 405.
Available
http://www.nature.com/nature/journal/v405/n6782/abs/405037a0.html

Eight more Fukushima kids found with thyroid cancer; disaster link denied. (2014,
February 7). *The Japan Times.* Available
http://www.japantimes.co.jp/news/2014/02/07/national/eight-more-fukushima-
kids-found-with-thyroid-cancer-disaster-link-denied/#.U2Zvr61dVXo

EX-SKF. (2014, February 13). (UPDATED) (Now they tell us) highly radioactive
pieces found in Naraha-Machi in June/July 2013 came from #Fukushima I
NPP, TEPCO now says. Available http://ex-skf.blogspot.com.ezproxy1.lib.
asu.edu/2014/02/now-they-tell-us-highly-radioactive.html

Forster, L., Forster, P., Lutz-Bonengel, S., & Brinkmann, B. (2002). Natural
radioactivity and human mitochondrial DNA mutations. *PNAS, 99*(21), 13950-
13954.

Fucic, A., Brunborg, G., Lasan, R., Jezek, D., Kundsen, L. E., & Merlo, D. F.
(2008). Genomic damage in children accidentally exposed to ionizing
radiation: A review of the literature. *Mutation Research, 658*(1-2), 111-123.

Fujisaki, M., & Ebuchi, T. (2013, December 17). Plan in works to ease TEPCO's
cleanup burden. *The Asahi Shimbun.* Available
http://ajw.asahi.com/article/0311disaster/fukushima/AJ201312170046

Fujiwara, S. (2014, February 25). Health risk or not? Cesium levels high in
hundreds of Fukushima reservoirs. *The Asahi Shimbun.* Available
http://ajw.asahi.com/article/0311disaster/fukushima/AJ201402250071

Fukushima Independent Investigation Commission. (2012). *The official report of
the Fukushima nuclear accident independent investigation commission.* Japan:

Japan National Diet. Available http://naiic.go.jp/wp-content/uploads/2012/07/NAIIC_report_hi_res2.pdf

Gerber, M. E., Reilly, B. K., Bhayani, M. K., Faust, R. A., Talavera, F., Sadeghi, N. & Meyers, A. D. (2013). Pediatric thyroid cancer. Available http://emedicine.medscape.com.ezproxy1.lib.asu.edu/article/853737-overview

Goliszek, A. (2003). In the name of science: A history of secret programs, medical research, and human experimentation. New York City: St. Martin's Press.

Gould, J. M., & Goldman, B. A. (1991). *Deadly deceit: Low-level radiation high-level cover-up*. New York: Thunder's Mouth Press.

Govt report to IAEA suggests situation worse than meltdown. Melt-through' at Fukushima? (2011, June 8). *The Daily Yomiuri Online.* Available http://fukushimanewsresearch.wordpress.com.ezproxy1.lib.asu.edu/2011/06/08/japan-melt-through-at-fukushima-govt-report-to-iaea-suggests-situation-worse-than-meltdown/

Government without oversight. (2013, December 6). *The Japan Times* Available http://www.japantimes.co.jp/opinion/2013/12/06/editorials/government-without-oversight/#.UqN_vuLlfzM

Hall, K., & Williams, C. J. (2011, March 15). Fire erupts again at Fukushima Daiichi's no. 4 reactor; nuclear fuel rods damaged at other reactors. *Los Angeles Times.* Available http://articles.latimes.com/2011/mar/15/world/la-fgw-japan-quake-reactor-fire-20110316

Hasegawa, K. (2013, July 24). Fukushima nuclear clean-up to cost $58 bn. *Phys.Org.* Available http://phys.org/news/2013-07-fukushima-nuclear-clean-up-bn.html

Haworth, A. (2013, February 23). After Fukushima: Families on edge of meltdown. *The Guardian.* Available http://www.theguardian.com/environment/2013/feb/24/divorce-after-fukushima-nuclear-disaster

Hayashi, Y., & Morse, A. (2011, March 16). Setback in the reactor fight. *The Wall Street Journal*, p. A1.

How radiation affects cells. (2007). Available http://www.rerf.jp/radefx/basickno_e/radcell.htm

Institute of Nuclear Power Operations. (2011) *Special report on the nuclear accident at the Fukushima Daiichi nuclear power station.* Available http://www.nei.org/resourcesandstats/documentlibrary/safetyandsecurity/reports/special-report-on-the-nuclear-accident-at-the-fukushima-daiichi-nuclear-power-station

Jaeschke, B. C., & Bradshaw, C. (2013). Bioaccumulation of tritiated water in phytoplankton and trophic transfer of organically bound tritium to the blue mussel, mytilus edulis. *Journal of Environmental Radioactivity, 115,* 28-33.

Japan groups alarmed by radioactive soil. (2011, July 5). *Agence France-Presse.* Available http://www.google.com.ezproxy1.lib.asu.edu/hostednews/afp/ article/ALeqM5ivr747xKaxw9RGq5zMSDO-On_WRQ?docId=CNG. 2875ee35cc28aa30725ee1bfd4cfbde.111

Kan, N. (2013, October 28). Encountering the Fukushima Daiichi accident. *The Huffington Post.* Available http://www.huffingtonpost.com/naoto-kan/japan-nuclear-energy_b_4171073.html

Kasai, T. (2013, March 11). About 60 percent of Fukushima evacuees cannot return home by 2017. *The Asahi Shimbun.* Available http://ajw.asahi.com/article/0311disaster/recovery/AJ201303110005

Kihara, T., & Aoki, M. (2013, January 17). Crooked cleanup: Photos, videos show contractors lied in decontamination reports. *The Asahi Shimbun.* Available http://ajw.asahi.com/article/0311disaster/fukushima/AJ201301170063

Kimura, S., & Hattori, H. (2014, January 31). Debris hinders decommissioning work at Fukushima nuclear plant. *The Asahi Shimbun.* Available http://ajw.asahi.com/article/0311disaster/fukushima/AJ201401310216

Kirsch, S. (2004). Harold Knapp and the geography of normal controversy: Radioiodine in the historical environment. *Osiris, 19*, 167-181.

Körblein, A., & Küchenhoff, H. (1997). Perinatal mortality in Germany following the Chernobyl accident. *Radiation and Environmental Biophysics, 36*, 3-7. Available http://www.alfred-koerblein.de/chernobyl/downloads/ KoKu1997. pdf

Kumano, Y. (2010). *Integrity inspection of dry storage casks and spent fuels at Fukushima Daiichi nuclear power station* (PowerPoint slides, PDF Document). Available http://www.nirs.org/reactorwatch/accidents/6-1_powerpoint.pdf

Kusunoki, Y., Yamaoka, M., Kubo, Y., Hayashi, T., Kasagi, F., Douple, E. B., & Nakachi, K. (2010). T-cell immunosenescence and inflammatory response in atomic bomb survivors. *Radiation Research, 174*(6), 870-876.

Lower radiation readings proposed to speed return of Fukushima evacuees (2013, November 9). *The Asahi Shimbun.* Available http://ajw.asahi.com/article/behind_news/social_affairs/AJ201311090063

Lutz-Bonengel, S., Brinkmann, B., Forster, L., Forster, P. & Willkomm, H. (2002). Natural radioactivity and human mitochondrial DNA mutations. *Proceedings of the National Academy of Sciences of the United States of America, 99*(21). Available http://www.pnas.org/content/99/21/13950.long

McCurry, J. (2011, June 8). Fukushima nuclear plant may have suffered "melt-through," Japan admits. *The Guardian.* Available http://www.theguardian.com/world/2011/jun/08/fukushima-nuclear-plant-melt-through?CMP=twt_gu

Ministry of the Environment (2013). *Progress on off-site cleanup efforts in Japan* (PowerPoint slides, PDF Document). Available http://www.iaea.org/newscenter/news/2013/cleanup160913.pdf

Mochuzuki, I. (2014, January 12). Tepco "There is no way to shield bremsstrahlung from contaminated water tanks." *Fukushima Diary*. Available http://fukushima-diary.com/2014/01/tepco-there-is-no-way-to-shield-bremsstrahlung-from-contaminated-water-tanks/

Møller, A. P., Bonisoli-Alquati, A., Rudolfsen, G., & Mousseau, T. A. (2011). Chernobyl birds have smaller brains. *PlOS One, 6*(2), e16862.

Nader, R. (2014, January 24). The Fukushima secrecy syndrome – from Japan to America. *Common Dreams*. Available https://www.commondreams.org/view/2014/01/24-8

Nadesan, M. (2013). *Fukushima and the privatization of risk*. Houndmills, UK: Palgrave Pivot.

Nagata, K. (2013, August 20). TEPCO yet to track groundwater paths. Liquefaction threat adds to Fukushima ills. *The Japan Times*. Available http://www.japantimes.co.jp/news/2013/08/20/national/tepco-yet-to-track-groundwater-paths/#.U2XHpF7K3yi

Nagata, K. (2014, March 6). Solving Fukushima water problem a long, hard slog. *The Japan Times*. Available http://www.japantimes.co.jp/news/2014/03/06/national/solving-fukushima-water-problem-a-long-hard-slog/#.U2XIE17K3yh

Nagata, K. (2014, January 15). TEPCO business plan, including July reactor restarts, gets official OK. *The Japan Times*. Available http://www.japantimes.co.jp/news/2014/01/15/national/tepco-business-plan-including-july-reactor-restarts-gets-official-ok/#.UtalOBBdX0c

Negishi, T., Ozawa, K., Seino, Y., & Otsuki, N. (2013, December 27). TEPCO to pay evacuees additional 7 million yen for "loss of hometowns." *The Asahi Shimbun*. Available http://ajw.asahi.com/article/0311disaster/fukushima/AJ201312270055

Ness, K. K., Krull, K. R., Jones, K. E., Mulrooney, D. A., Armstrong, G. T., Green, D. M., & Hudson, M. M. (2013). Physiologic frailty as a sign of accelerated aging among adult survivors of childhood cancer: A report from the St. Jude lifetime cohort study. *Journal of Clinical Oncology, 52*. Available http://jco.ascopubs.org.ezproxy1.lib.asu.edu/content/early/2013/11/18/JCO.2013.52.2268.abstract

Nesterenko, A. V., Nesterenko, V. B., & Yablokov, A. V. (2009). Introduction: The difficult truth about Chernobyl. *Chernobyl: Consequences of the catastrophe for people and the environment* (pp. 1-3). Boston, Massachusetts: Blackwell Publishing.

NHK radio regular quits after anti-nuclear commentary nixed. (2014, January 30). *The Asahi Shimbun*. Available http://ajw.asahi.com/article/behind_news/social_affairs/AJ201401300075

Nishikawa, J. (2014, January 24). Scientists: Cosmic rays can see through nuclear reactor, locate fuel. *The Asahi Shimbun.* Available http://ajw.asahi.com/article/0311disaster/fukushima/AJ201401240069

Nogami, H., & Fujisaki, M. (2013, December 16). Banks look to cut off new lending to TEPCO after 300 billion yen loans. *The Asahi Shimbun.* Available http://ajw.asahi.com/article/0311disaster/fukushima/AJ201312160068

Nose, T., & Oiwa, Y. (2014, February 8). Thyroid cancer cases increase among young people in Fukushima. *The Asahi Shimbun.* Available http://ajw.asahi.com/article/0311disaster/fukushima/AJ201402080047

Nuclear accident disclosure. (2011, July 8). *The Japan Times.* Available http://search.japantimes.co.jp/cgi-bin/ed20110708a1.html

Nuclear plant restarts on the table. (2014, January 12). *The Japan Times.* Available http://www.japantimes.co.jp/opinion/2014/01/12/editorials/nuclear-plant-restarts-on-the-table/#at_pco=tcb-1.0&at_tot=8&at_ab=-&at_pos=3

Nuclear power plant insurance problem [Genpatsu Hoken Mondai: Tōden 1,200-oku En Kyōtaku Shōnin]. (2012, June 13). *Tokyo Shimbun.* Available http://www.tokyo-np.co.jp/article/feature/nucerror/list/CK2012011302100010.html

OECD (2001). Glossary of statistical terms: environmental externalities. Available http://stats.oecd.org/glossary/detail.asp?ID=824

Ohira, K., & Fujisaki, M. (2012, July 31). Taxpayers, electricity users finance TEPCO bailout. *The Asahi Shimbun.* Available http://ajw.asahi.com/article/0311disaster/fukushima/AJ201207310068

Ohmae, K. (2011, October 28). *Lessons of Fukushima Dai-ichi.* Available http://pr.bbt757.com/eng/pdf/finalrepo_111225.pdf

Okuyama, T., & Sunaoshi, H. (2013, December 17). State secrets law raises concern about safety of nuclear power plants. *The Asahi Shimbun.* Available http://ajw.asahi.com/article/0311disaster/fukushima/AJ201312170006

Osaka, E. (2012). Corporate liability, government liability, and the Fukushima nuclear disaster. *Pacific Rim Law & Policy Journal, 21*(3), 433-459. Available http://digital.law.washington.edu/dspace-law/bitstream/handle/1773.1/1161/21PRPLJ433.pdf?sequence=1

Panel willing to extend compensation period for Fukushima evacuees. (2013, October 26). *The Asahi Shimbun.* Available http://ajw.asahi.com/article/0311disaster/fukushima/AJ201310260046

Perrow, C. (2013). Nuclear denial: From Hiroshima to Fukushima. *The Bulletin of the Atomic Scientists, 69*(5), 56-67.

Priyadarshi, A., Dominguez, G., & Thiemens, M. H. (2011). Evidence of neutron leakage at the Fukushima nuclear plant from measurements of radioactive[35]S in California. *Proceedings of the National Academy of Sciences, 108*(35), 14422-14425.

Record cesium level found in groundwater beneath Fukushima levee. (2014, February 14). *The Asahi Shimbun.* Available http://ajw.asahi.com/article/0311disaster/fukushima/AJ201402140041

Record outdoor radiation level detected at Fukushima plant. (2013, December 7). *The Asahi Shimbun.* Available http://ajw.asahi.com/article/0311disaster/fukushima/AJ201312070041

Record strontium-90 level in Fukushima groundwater sample last July. (2014, February 7). *The Japan Times.* Available http://www.japantimes.co.jp/news/2014/02/07/national/record-strontium-90-level-in-fukushima-groundwater-sample-last-july/#.U2XIw17K3yh

Request for Compensation (2013, March 16). *The Nikkei Shimbun.* Available http://www.nikkei.com/article/DGXNASDD1600S_W3A110C1EB2000/

Ryall, J. (2011, June 9). Nuclear fuel has melted through base of Fukushima plant. *The Telegraph.* Available http://www.telegraph.co.uk/news/8565020/Nuclear-fuel-has-melted-through-base-of-Fukushima-plant.html

Ryall, J. (2014, March 3). Prosecutors drop charges over Fukushima nuclear disaster. *The Telegraph.* Available http://www.telegraph.co.uk/news/worldnews/asia/japan/10672216/Prosecutors-drop-charges-over-Fukushima-nuclear-disaster.html

Secret designation clause Fukushima IAEA, Fukui share private information. (2013, December 31). *The Tokyo Shimbun.* Available http://www.tokyo-np.co.jp/article/national/news/CK2013123102000114.html

Sokai, S. (2013, April 25). Conclusion of the judgment by Sendai high court and remarks by supporters. Available http://fukushima-evacuation-e.blogspot.com.ezproxy1.lib.asu.edu/2013/04/conclusion-of-judgment-by-sendai-high.html

Starr, S. (2012, February 16). Health threat from cesium-137. *The Japan Times.* Available http://www.japantimes.co.jp/text/rc20120216a1.html

Suga, M., Okada, Y., & Adelman, J. (2014, February 20). TEPCO finds new leak of radioactive water at Fukushima site. *Bloomberg.* Available http://www.bloomberg.com/news/2014-02-20/tepco-says-new-leak-of-radioactive-water-found-at-fukushima-site.html

Suman, S., Rodriguez, O. C., Winters, T. A., Fornace Jr, A. J., Albanese, C., & Datta, K. (2013). Therapeutic and space radiation exposure of mouse brain causes impaired DNA repair response and premature senescence by chronic oxidant production. *Aging, 5*(8), 607-622.

TEPCO official says Fukushima plant situation "out of control." (2013, September 13). *Xinhuanet.* Available http://news.xinhuanet.com/english/world/2013-09/13/c_125386760.htm

TEPCO seeks more government support as Fukushima costs soar. (2012, November 7). *The Asahi Shimbun.* Available http://ajw.asahi.com/article/0311disaster/fukushima/AJ201211070086

TEPCO to review erroneous radiation data. (2014, February 9). *NHK Online.* Available http://www3.nhk.or.jp/nhkworld/english/news/20140209_80.html

TEPCO withheld Fukushima radioactive water measurements for 6 months. (2014, January 9). *The Asahi Shimbun.* Available http://ajw.asahi.com/article/0311disaster/fukushima/AJ201401090060

TEPCO's post-mortem shows no. 2 reactor main source of radiation. (2012, May 25). *The Asahi Shimbun.* Available http://ajw.asahi.com/article/0311disaster/fukushima/AJ201205250053

The Associated Press. (2013, April 25). Japan court rejects demand to evacuate Fukushima children. *The Asahi Shimbun.* Available http://ajw.asahi.com/article/0311disaster/fukushima/AJ201304250125

Tokonammi, S., Hosoda, M., Akiba, S., Sorimachi, A., Kashiwakura, I., & Balonov, M. (2012). Thyroid doses for evacuees from the Fukushima nuclear accident. *Scientific Reports, 2*(507) doi:10.1038/srep00507.

Tonkin, S. (2011, May 18). Japan government draws lessons from March quake, pledges greater transparency. *World Economic Forum.* Available http://www.weforum.org/news/japan-government-draws-lessons-march-quake-pledges-greater-transparency

Trani, D., Datta, K., Dorion, K., Kallakury, B., & Fornace Jr, A. J. (2010). Enhanced intestinal tumor multiplicity and grade in vivo after HZE exposure: Mouse models for space radiation risk estimates. *Radiation and Environmental Biophysics, 49*(3), 389-396.

Uranaka, T. (2012, March 8). Japan lenders ready to back Fukushima operator TEPCO, but wary. *Reuters.* Available http://www.reuters.com/article/2012/03/08/us-japan-tepco-banks-idUSBRE8270EF20120308

Varma, S. (2014, February 14). Fukushima radiation data is wildly wrong, management apologizes. *The Times of India.* Available http://timesofindia.indiatimes.com/world/rest-of-world/Fukushima-radiation-data-is-wildly-wrong-management-apologizes/articleshow/30163944.cms?referral=PM

Vásquez-Maignan, X. (2011). Fukushima liability and compensation. *Nuclear Law Bulletin, 2*(88), 61-64.

Wertelecki, W. (2010). Malformations in a Chernobyl-impacted region. *Pediatrics, 125*(4), e836-e843. doi:10.1542/peds.2009-2219

Wetherbee, G. A., Debey, T. M., Nilles, M. A., Lehmann, M. B., & Gay, D. A. (2012). Fission products in national atmospheric deposition program—wet deposition samples prior to and following the Fukushima Dai-ichi nuclear power plant incident, March 8–April 5, 2011. *U.S. Geological Survey, 1277,* 6.

Writer calls for change in secretive bureaucratic culture. (2014, January 7). *The Mainichi.* Available http://mainichi.jp/english/english/newsselect/news/20140107p2a00m0na005000c.html

Xu-Friedman, M. A., & Regehr, W. G. (1999). Presynaptic strontium dynamics and synaptic transmission. *Biophysical Journal, 76*(4), 2029-2042.

Yamaguchi, M. (2013, November 26). Japan's secrecy law stirs fear of limits on freedoms. *Associated Press.* Available http://hosted.ap.org.ezproxy1.lib.asu.edu/dynamic/stories/A/AS_JAPAN_SEC RECY_LAW?SITE=AP&SECTION=HOME&TEMPLATE=DEFAULT

Zhang, W., Friese, J., & Ungar, K. (2013). The ambient gamma dose-rate and the inventory of fission products estimations with the soil samples collected at Canadian embassy in Tokyo during Fukushima nuclear accident. *Journal of Radioanalytical and Nuclear Chemistry, 296*(1), 69-73.

Zheng, J., Tagami, K., Watanabe, Y., Uchida, S., Aono, T., Ishii, N., & ...Ihara, S. (2012). Isotopic evidence of plutonium release into the environment from the Fukushima DNPP accident. *Scientific Reports,2* doi:10.1038/srep00304.

Notes

i. The press release has since been removed from the IAEA webpage. It was originally found here http://www.iaea.org/press/?p=1248.

ii. The press release has been removed from the IAEA webpage. It was originally found here http://www.iaea.org/press/?p=1252.

iii. For example, The Asahi Shimbun reported "TEPCO announced Record cesium level found in groundwater beneath Fukushima levee" February 14, 2014 (link http://ajw.asahi.com/article/0311disaster/fukushima/AJ201402140041). The article said that cesium found in groundwater under a coastal levee near unit 1 spiked from 76,000 Becquerels per liter on February 12, 2014 to 130,000 Becquerels per liter on February 13, reaching the highest level of cesium ever detected at that location.

iv. However, the study did specifically note that children who had acute lymphoblastic leukemia who were treated with high doses of radiation to the brain showed signs as adults of brain changes and memory problems compared to adults in their age group.

ENFORCED AMNESIA – THE DISPOSSESSION OF MEMORY

Paul Langley

Victors write the history books. The thrust of the official view washes over individual experience and memory, threatening to submerge it. The vitality of insight memory brings is lost if one stops swimming against the tide. The counter-current of the official view bends the wind and it whips the face of the solitary witness with the cold cutting chill of denial.

He had seen the Black Rain. Living in Hiroshima, he suffered as many did in those suburbs, washed by the tar-like liquid that fell from the bomb cloud in 1945. That Black Rain fell far more widely than admitted at that time or since. Now even his own government doubted that his suffering was real. Bent and weakened, often bed-ridden, the years had been an ordeal. Bura Bura disease. Ha. Too much worrying about nothing, the nation's experts had said recently. His skin had borne witness to the truth.

Figure 1 Trousers with black rain stains
Location: Koi-machi (now Koi-ue). Distance from hypocenter: approx 3,700m.
When the team traveled to west Hiroshima to study the effects of black rain, they discovered these stained trousers in a private home in Koi-machi and learned that the rain was like dirty water. Photograph by Shigeo Hayashi. Courtesy of Hiroshima Peace Memorial Museum. Copyright image, reproduced with permission of the Hiroshima Peace Memorial Museum.

The day the smoke went up at Fukushima Daiichi, the old man's children remembered, and knew that their own "Black Rain" was on its way. They knew what they had to do, for it would last a long time. First: Remember. Don't let there be forgetfulness. Build a living library of facts sufficient to silence the voices of denial. The nation's memory extends beyond the horizon. It will always be August 1945. It will always be March 2011. Few remain with recollection of the Black Rain as we watch Fukushima unfold.

Figure 2 Black Rain, Hiroshima 1945; radioactive substances on a roof
Location: Koi-machi (now Koi-ue). Distance from hypocenter: approx. 3,700m.
Japanese scientists climbed on a roof in Koi-machi to collect soil and mud for samples to measure residual radiation. They looked for gutters, cracks between roof tiles, or anywhere they might find a puddle of water. Photograph by Shigeo Hayashi. Courtesy of Hiroshima Peace Memorial Museum. Copyright image, reproduced with permission of the Hiroshima Peace Memorial Museum.

The children stood ready to demolish the cruel official claim that they were nothing but fearful worriers. They held their courage high. With calm resolution they formed a blockade against the official attempt to impose old lies onto the latest disaster.

Fukushima

On the 15th of March 2011, the words of the Chief Scientist of Britain, Lord Beddington, were transmitted and streamed around the world by the BBC. I watched him speak on local TV in Adelaide. If there were a meltdown, he said, "you would get an explosion and radioactive material would be emitted. But it would be emitted to about 500 meters and it would be a relatively short duration of the order of an hour or so. Compare that with Chernobyl…"[1]

Not long after Beddington spoke, the Prime Minister of Japan, dressed in his blue overalls, appeared at the rostrum in Japan. He pleaded for calm. The situation was urgent. He and his Cabinet Secretary, Mr. Edano, told Japan and the world that radiation from the reactor plant was "now high enough to endanger health." People up to 10km outside of the 20km evacuation zone were told to stay indoors with doors and windows shut. Their homes became their "containment domes". 250km to the south, Tokyo was receiving far higher levels of radiation than normal. The people there were told they were perfectly safe. China responded by evacuating its citizens from Japan.[2]

In 1967, the US Atomic Energy Commission ordered a study into the consequences of cooling failure and meltdown in power reactors.[3] This study by the Ergen Committee found that the nature of the strata beneath a failed reactor in meltdown affected the progress of the disaster. This report describes the expansion of molten corium in part as follows:

> To illustrate the magnitude of the decay heat release from a 3,200 MWt core in terms of the heat capacity of several common materials, the size of a molten sphere containing one-hour-aged fission products was calculated using dry sand and limestone as the heat absorbing material. These results indicate that molten spheres of approx 60 ft diameter for limestone and approx 90 ft diameter for dry sand would be required to absorb and dissipate the decay heat from a 3,200 MWt core. Also the growth of a molten sphere would continue for approx 20,000 hours under these conditions.[3]

The production of vast amounts of highly contaminated water in the event of a major accident at Fukushima Daiichi was a hazard likely to be known about prior to the start of construction of Unit 1 in 1967.[4][5] I have reason to think that the threat to the ocean and to the fisheries off the Japanese coast has long been known to the nuclear industry in the event of a major accident occurring at any Japanese reactor site. I also wonder whether Mr. Ergen and his consulting companies, General Electric and Westinghouse, considered wet sand advantageous in the attempt to mitigate meltdown consequences.

The following June 25, 2012 radio report by the Australian Broadcasting Corporation was predictable by the facts of history:

> Fukushima radiation kills fishing industry. Fishermen in the Fukushima area, the site of the greatest single radiation contamination of the ocean in history, fear their industry is ruined. While boats still go to sea, their catches are being withheld from sale and instead the fish is sent for analysis. And it's not just Japanese fishermen recording radioactive hauls. Contamination is turning up thousands of kilometres away in fish caught on the other side of the Pacific Ocean.[6]

However, it was not until August 2013 that nuclear authorities in Japan and around the world admitted what they had known since March 2011. The broken reactors had been leaking radioactive pollution into the ground water and hence into the sea since the start of the disaster.[7] The venting into air and emissions into water lasted longer than Beddington's magic hour. It is still going on, and will continue for more years to come. The Ergen, 1967, study shows that at 100,000 hours after meltdown, the fuel is still molten.[3]

With the loss of the power grid and emergency diesel generators, a power pole becomes a symbolic crucifix for ordinary people and their families. The nuclear industry points with pride to design achievements at Fukushima even today. For instance, in regard to the explosive failure of reactor 2, the American Nuclear Society states the following: "[the] RCIC system operated for ~70 hours. In general, one should not expect the RCIC system to run much beyond 8 hours in a station blackout"[8] (RCIC – Reactor Core Isolation Cooling system – an integral emergency cooling system). The frantically scribbled log the engineers kept on a whiteboard in the control room as the nuclear plant slid towards disaster reveals the chaos of those 70 hours: "15:42, nuclear emergency declared. 15:50, loss of water level readings. 16:36, emergency core cooling system malfunction. No water can be injected."[9]

The "designed-in" roots of the failure of reactor emergency core cooling systems were described in the 1970s, as explained by Nader and Abbott in their 1977 book, *The Menace of Atomic Energy*. This book devotes a section to the flaws known to be inherent in reactor Emergency Core Cooling Systems at the time of its publication.[10] In the mid-1970s, even the Nuclear Regulatory Commission's own consultants held the view that the emergency systems built into reactors to prevent core overheating were unreliable. Keith Miller, Professor of Mathematics at the University of California, Berkley stated that the emergency systems were "no more reliable than the weather". Miller called for a halt to the licensing of nuclear reactors.[10] In addition to Miller, Nader and Abbott were to provide a list of another 10 expert individuals and organisations, many employed by the US nuclear regulator, concerned about the adequacy of reactor emergency cooling systems, including: Dr. Morris Rosen, technical advisor to the Director of reactor licensing at the Atomic Energy Commission; George Brockett, a leading ECCS researcher certified by the Atomic Energy Commission; J. Curtis Haire, person in charge of the AEC research effort on reactor emergency cooling systems (J. Curtis Haire accused the AEC of censuring his reports); Milton Shaw, head of the US civilian nuclear power program; Alvin M. Weinberg, director of the Oak Ridge National Laboratory; the Reactor Safety Committee of the Federal Republic of Germany; The Federation of American Scientists; the RAND Corporation; the AEC Advisory Committee on Science and Technology; the investigation committee examining ECCSs for the Swedish government; and the Pugwash Conference.[10]

One of the failures of the nuclear industry and mass media narrative since the March 2011 disaster has been its enforced amnesia in regard to the warnings the industry and regulators have had since the probability of core melt in the event of cooling system and emergency cooling system failure was first recognised in the 1960s. It has been over 40 years since these warnings were issued and ignored. Nothing has changed. The enormity of the natural disaster did not cause the inadequacy of the Fukushima Daiichi emergency cooling systems. **The designed-in inadequacy of reactor ECCSs have been known by many industry experts for decades. While these issues do relate specifically to the Fukushima**

Daiichi General Electric type reactors in this instance, they also relate to ALL types of commercial power reactors designed and produced by the United States at that time and since. As late as 1985, a former Chair of the US Atomic Energy Commission was so concerned that he called for a "technical fix": "David Lilienthal, the first Chairman of the Atomic Energy Commission … called on nuclear engineers to come up with a technical fix: a reactor that was transparently and patently immune from a core melt. This he regarded as the key to a rebirth of nuclear."[11] The call for a "technical fix" obviously remains unanswered in 2014. The culpability of the industry is seen in the foresight it has had for DECADES.

During the 1970s, the perilous state of multi-megawatt nuclear reactors was so well known among technologists and to some members of the public that demand for popular explanations of the findings of the Ergen Report was high at the time. In 1971, nuclear scientist Ralph Lapp wrote an article that was published in *The New York Times*. In this article, Lapp provides a synopsis of the Ergen Report, explaining that in dry sand a molten core "might persist for a decade. This behaviour projection is known as the China Syndrome…. Utilities should curb their plans to site nuclear plants near cities…."[12] Where the late Dr. Lapp would site nuclear reactors is unknown given today's knowledge of environmental interaction, the scarcity of land, the inability of Japan to feed itself in the best of times, the experiences of Mayak, Windscale, Three Mile Island, Chernobyl and Fukushima Daiichi. His 1971 article "Thoughts on Nuclear Plumbing" contains a tragically prophetic set of observations about a nuclear industry intent on cutting corners and minimising the public perception of inherent dangers. In the case of Fukushima Daiichi, the problems of nuclear plumbing are worsened by the decision to locate the reactors over a high flow rate aquifer.[13] [14]

An aquifer may provide corium cooling – though Ergen covers only dry sand, not wet sand – but such a setting is nothing more than a radionuclide dispenser and distributor, the existence of which at Daiichi was hidden for two years by world nuclear authorities, including the Tokyo Electric Power Co. (TEPCO) and the Japanese government.[15] The flow of radionuclides offsite into the coastal waters had been denied for many months by TEPCO and Japanese government authorities.[16]

Whatever the actual motivation for placing the Fukushima Daiichi plant over such an aquifer, TEPCO and the rest of the Japanese and world nuclear industry has had since 1967 to consider and design a system for catching, holding and decontaminating the contaminated water caused by meltdowns. The Ergen Report of 1967 shows that nuclear authorities have considered meltdown a possibility, **from whatever cause**, in any multi-megawatt reactor built since that time.

The constant denials by nuclear authorities in regard to the constant emissions by the failed Fukushima reactors came in spite of the routine findings to the contrary by those independent researchers who persisted with sampling, despite the flood of industry propaganda. Ken Buesseler, a senior scientist with the Woods Hole Oceanographic Institution, has monitored thousands of fish taken from the ocean off Fukushima and has found continuing high levels of short lived Cesium-134 in the ocean life. Buesseler stated "It's getting into the ocean, no doubt about it," he said. "The only news was that they finally admitted to this."[17]

By the 15th of March 2011, more than 200,000 people had been evacuated from the exclusion zone around the crippled plant. Authorities ordered the evacuations.[18] If the evacuations were adequate, if the exclusion zone was realistic and appropriate, why were people who were located *outside the exclusion zone* ordered to stay indoors, with no prior warning, from March 15, 2011?[2] Many had to stay indoors, isolated, for considerable periods of time. The nuclear experts, if one is to believe them, claim to know all there is to know about safely predicting consequences. Beddington's statements of March 15, 2011 are revealed to be what they are – a complete crock – by the order to stay indoors at a range of 30 km, issued on the same day as his propaganda.[1][2] Further evidence of duplicity is found in samples of Japanese fauna and flora. In the mountains of Japan, 400 kilometres away from Fukushima, scientists go fishing with Geiger counters in order to measure the risk posed by freshwater fish high in the mountains. They have mapped where the Fukushima rain and dust falls.[19] It meets the horizon and it follows the arc of the earth.[20]

Having failed to contain the fallout, nuclear authorities attempt to contain the thoughts of populations. They try to construct denial within the collective awareness with their platitudes:

> There is no credible risk of a serious accident . . . the risk of meltdown is extremely small . . . . Those spreading FUD [fear, uncertainty and doubt] at the moment will be the ones left with egg on their faces. I am happy to be quoted forever after on the above if I am wrong . . . but I won't be. The only reactor that has a small probability of being "finished" is unit one. And I doubt that, but it may be offline for a year or more.

This is a statement from an "expert" employed by Adelaide University, South Australia, March 2011, as quoted over the national broadcaster in analysis as the disaster unfolded.[21] In those early days, the world was awash with the voices of an army of nuclear King Canutes, attempting, it seems, to hold back the flood of nuclear pollution from the public awareness. They failed to hold back the tide of truth as experienced by the people of Japan. It is no longer 1945, there is no legitimate basis for censorship, the people will be heard.

Hiroshi Sano spoke of his family's ordeal in Iitate Village, 20kilometers outside the evacuation zone. The village was subject to heavy fallout from the reactors, but for an entire month no official told the village. The farmland lies ruined by fallout from the nuclear plant. The people have become scattered nuclear refugees: "I remember a stream of evacuees coming from the direction of the disaster. I never imagined that I myself would have to evacuate." Sano says Iitate's residents were becoming more fearful even before the evacuation order, as bulletins from the International Atomic Energy Agency and rumours on the Internet (which turned out to be correct) suggested the town had been showered with wind-borne fallout.[22] The withholding of computer-generated fallout prediction information from the nation's "SPEEDI" system is now infamous worldwide. Japan gave the US military the data, and withheld it from its own embattled and struggling people. Major newspapers in Japan describe the details of the "SPEEDI Deception", as it is known.[23]

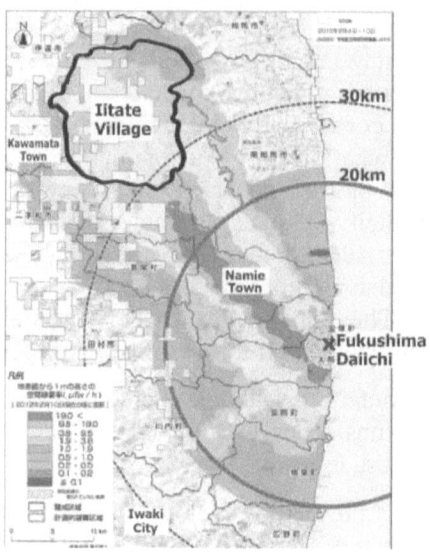

Figure 3 Map showing the location of Iitate Village in relation to the nuclear fallout of March 2011. *Based on the website of the Ministry of Education, Culture, Sports, Science and Technology*

In the end, the IAEA heard the month-long pleadings from the independent monitors from Greenpeace. The Japanese government was forced, by international pressure, into evacuating a village it knew all along to be dangerously contaminated by nuclear fallout.[24][25]

We remain possessed of our memories. No false recounting will cause our collective amnesia. The whole official history of the nuclear world is as a singular photograph of such disaster filtered by rosy recommendations issued in the absence of full disclosure by those who put the value of their status above the need for truth and openness in the face of severe adversity suffered by ordinary people who are reliant upon those arrogant authorities.

Understanding Fukushima involves an understanding of the past as well as the present. I aim in this paper to walk from August 1945 until March 2011, trying to discover the themes that define the individual and official responses to disaster. Is the withholding of truth, so much in evidence since March 2011, unique in nuclear history? Or has the technology repeatedly presented peoples with very large costs to bear? Costs held as secrets until it was too late to escape them?

Japanese scientist Yoshinobu Masuda is among those who consider that the survivors of the radioactive dust and Black Rain of Hiroshima in August 1945 bear witness to a suppression of facts of great importance in the aftermath of the Fukushima Daiichi disaster.[26] Dr. Shuntaro Hida is one of the doctors who witnessed the Hiroshima atomic bomb. He is among those who describe a disease syndrome discounted by western health physics. This disease blighted the lives of many Japanese people in the wake of August 1945; survivors suffer it still today. In Japan is it called Bura Bura disease. It is a chronic multi-system disorder.[27] A fuller understanding of radiation effects, including phenomenon such as Bura Bura disease, demands that we see Fukushima Daiichi as part of a historic process, as well as a specific event. We have to travel back to 1945 in order to gain the context within which crucial truths seem to have been suppressed and buried beneath the myth of nuclear progress.

The First American into Nagasaki and the Long March to the Truth

Dr. Nello Pace was the first American to enter Nagasaki. He had been ordered there by an outraged General Douglas McArthur. McArthur had not been advised in advance of the atomic bombing of Japan. He thundered: "I want my people, my doctors, in there looking at the effects of this, not these Manhattan District people!" So it was that Dr. Pace entered and surveyed Nagasaki, and then Hiroshima, without the clearance of General Groves and the Manhattan Project. Dr Pace tells the story:

> I wrote the reports, and I classified them Secret – which was as high as I was cleared for, and that was fine to me. So one day I got a phone call to go down and report to Ross McIntyre, the Surgeon General. I thought, "Maybe they're going to give me a medal....there's McIntyre sitting there and another guy that looked familiar; I think he was introduced to me as General Groves. They were there with a big scowl on their faces. They wanted to know why I'd classified this report so low, when it should have been classified Top Secret. I said, "I wasn't cleared for Top Secret. I had no idea."

> They said, "You should have classified it as Top Secret." This was a very serious thing. McIntyre said, "I want you to destroy any piece of paper that had anything to do with this, any of the data." "And furthermore," he said,

"I want you to forget. I'm ordering you to forget what you wrote in this report."[28]

The Manhattan Project did preserve the work. The integrity of Dr. Nello Pace is not in question here. The order he reports being given is indicative of the cultural norms of the time.

Dr. Pace and other US surveyors determined that both Hiroshima and Nagasaki were safe. However, Occupation troops from the US and Australia found later to their horror, distress and loss that this was not the case: The Australian Atomic ExServicemen's Association states that 90 percent of those Australians who served as Occupation troops within the British Commonwealth Occupation Force suffered cancers or otherwise died before their time due to their service as Occupation troops in Japan from 1946. The Australians were assigned to Hiroshima and Kure. The remaining survivors and widows who pursue justice despite the official claim Hiroshima was safe at the time have a hard battle. However, there have been some successes.[29] On 26 January 1989, the Australian Repatriation Commission awarded a pension to Mrs. Clements, after finding that the death of her husband, a veteran of the Occupation of Hiroshima, "arose out of . . . war service." The Commission accepted the evidence supported Mrs. Clement's claim that "the effects of radiation which (Mr. Clements) received by virtue of being stationed near Hiroshima for 15 months" and possibly other factors, led to the death of her husband from Hodgkin's Disease in 1954. Although Australian radiation protection officials state that all radioactive sources in Hiroshima had dissipated by 1946, multiple Australian courts, having considered the evidence, found this is not the case.[30] The Atomic ExServicemen's Association report documents the victories:

> Two widows beat the Commonwealth of Australia after it refused to grant them widow's pensions after their husbands died of cancer of the pancreas and Non-Hodgkin's Lymphoma related to post-war service in the ionising radiation contamination at Hiroshima, Japan. The lawyers for the women blasted the Repatriation Commission's eight year fight against the two as a disgraceful waste of taxpayer's money. Eight years ago (from 2000, the time of the original report) Mrs Theresa Connolly began her fight. The second widow, Mrs June Viola Flynn, died in 1998 (during the battle for justice.) The Administrative Appeals Tribunal on 26 March 1998 found both widows were entitled to war widows pensions, but the Repatriation Commission appealed against the decision. Australian Federal Court Justice Richard Cooper ordered costs against the Repatriation Commission.[31]

Even at the very late dates at which these widows received justice, we see the still-vital controversy that exists over the nature and extent of the radiation hazard remaining in the cities subjected to the 1945 atomic attacks during the subsequent periods.

What then were the consequences for those who grew up as children among the fallout and induced radioactivity? What is the story of the chronic exposure, the story which isn't told, as opposed to the one that is? We all know about acute radiation sickness. What about the chronic effects? Why is there argument about the boundaries of the Hiroshima Black Rain Area still today? For surely, where an area causes harm, there is a duty to properly map it. War or peace, this should have been done in 1945.

Three days after the destruction of Hiroshima, Manhattan Project Officer Dr. Robert Stone wrote two letters to Stafford Warren's deputy, and Stone's former student, Hymer Friedell. The second letter described Stone's "mixed feelings" at the success that had been achieved and his fear that the lingering effects of radiation from the bomb had been underestimated: "I could hardly believe my eyes," Stone wrote, "when I saw a series of news releases said to be quoting Oppenheimer, and giving the impression that there is no radioactive hazard. Apparently all things are relative."[32] The Franck Report warned the US government that the USA risked becoming an international pariah as a consequence of the atomic attacks. As a result of this, and out of fear of educating the USSR, the US, I believe, clammed up about radiation effects.[33]

In 1945, the entire world was at the receiving end of a news blackout that took decades to partially lift. And so even today, those who believe the American reports alone miss the additional evidence given by those who survived the bombs. Most importantly, the world misses the accounts given by those who, though out of range of the bomb blasts and the instantaneous radiation bursts of gamma and neutron rays, suffered from fallout exposure due to wet deposition. This is known in Japan as the Black Rain. The world missed the testimony of those who entered the bombed cities after the attacks and who suffered the consequences of the radioactive dust.

Since 1945, there has been a profound dichotomy between the announcements of nuclear officials and the experience of ordinary people. This is true of Japan, the Soviet Union and its captive states, and for the USA, Australia, Algeria, Polynesia – everywhere. Finding the truth in one place helps victims in the other places. Throughout this nuclear history, no Western authority admitted that chronic "low level" exposure caused any chronic health syndrome. But the reality of such a syndrome was first seen in Japan. The syndrome, named Bura Bura disease, has been diagnosed, treated and explained since 1945 by a Japanese doctor name Dr. Hida, among others.[27]

The Private Awareness of the Top Secret

On May 24, 2012, the Hiroshima Peace Media Centre reported the following research results concerning widespread and enduring contamination in Japan from the Hiroshima and Nagasaki bombings:

A team of researchers from Hiroshima University and other institutions, investigating traces of the "black rain" which fell in the aftermath of the atomic bombing of Hiroshima, has issued its findings. The results indicate that Cesium-137, a radioactive substance believed to be derived from the black rain, was found in soil beneath the floors of six sites, including residences in Yuki Town (Saeki Ward, Hiroshima City) and Akiota Town (Hiroshima Prefecture). These locations lie roughly nine to 22 kilometers from the bomb's hypocenter. This is the first time that traces of black rain have been confirmed in these areas, and all were found outside of the designated zone known as the "heavy rain area." Those within the "heavy rain area" at the time of the bombing are eligible for relief measures put in place by the central government. The research team, which includes Masayoshi Yamamoto, a professor in radiochemistry at Kanazawa University, investigated the soil under the floors of 20 homes built between 1946 and 1948. After digging to a depth of 30 centimeters, they discovered the traces of Cesium-137 in the soil.[34]

For many years, survivors have urged that Japanese authorities recognize that the Black Rain fell over a much wider area than is officially acknowledged. The City of Hiroshima recognizes that Black Rain fell over a wide, elliptically shaped, area. Hiroshima is surrounded by a teardrop-shaped area affected by Black Rain.[35]

The disparity between the Black Rain affected area as accepted by the Japanese government, and the area known to have suffered Black Rain in August 1945, is most important. The illnesses suffered by the survivors of the Atomic Bomb Black Rain are disabling and poorly documented in English-speaking nations. Japanese authorities acknowledge one group of sufferers, while another group is not recognized on the basis of an arbitrary demarcation. That criterion is irrational and cruel. The government response to radiation exposure since 1945 has thus also been irrational.

The Japanese government attempted to tighten the criteria by which Atomic Bomb survivors became eligible for medical and other assistance some years ago. The Citizens' Nuclear Information Centre of Japan reported in 2009 that Atomic Bomb survivors, having challenged the national government in court, had won 18 straight court cases. On June 9, 2009, the national government announced it would not challenge the court decisions and would review the criteria by which A-bomb disease sufferers are certified.[36] The government of Japan now appears determined to learn from its 2009 defeat in order to avoid suffering further setbacks in its battle against atomic bomb survivors, particularly those who are Black Rain afflicted.

Although it is now 68 years since the atomic bomb was detonated over Hiroshima, the full facts have yet to be acknowledged by government. Survivors still suffer, the true nature of the suffering is disputed and the administration of justice and care given to the survivors is subject to geographic whim on the part of government, rather than being based in scientifically established evidence. As we have seen, that evidence is still being uncovered in the medical histories and stories of past Black Rain survivors. Findings hold contemporary relevance for victims of 11 March 2011, despite "official" silence on the highly radioactive rain that fell in the early days after the disaster. The official history of Japan (as opposed to the experienced history lived and remembered by the survivors of August 1945), with its divergent view of past events, swamped coverage of the events at Fukushima Daiichi and its resultant fallout areas with pre-existing schema and scripts.

People have lived in hotspots in Japan since 1945. Since March 2011, people, including young children, have been forced to live in new hotspots created by the nuclear disaster of Fukushima Daiichi. The radiological history of Japan urgently needs specialist re-examination. The old lies are being imposed upon new, young, lives. I hope that the work of Masuda Yoshinodu and Dr. Hida will be acknowledged by government and used in deliberations regarding the fate of the people of Fukushima.[26] [27]

The Early Surveys of Hiroshima – the Lost "Snapshot"

Japanese science has responsibility, in its capacity as an adviser to government, to listen to the A-bomb witnesses. This is particularly important because witness accounts can now be examined in conjunction with Japanese surveys of Hiroshima that took place immediately after the bombing. These August 1945 Japanese surveys were conducted at great risk. However, the arrival of US troops, under direct command from the Manhattan Project's General Groves, and led by General Farrell in September 1945, resulted in the confiscation of the early Japanese survey data. Included were reports, measurements, samples – including soil samples and human remains. The Japanese scientists became subordinate to the American survey teams.[37][38][39] The confiscated samples and human remains were returned to Japan in the 1970s. This caused great celebration at the time in Japan. At last Japanese scientists could independently study the soil of their city taken in 1945 and assess what had occurred in August 1945 from a technical basis. Of course, the short lived and highly active fission and neutron induced substances were long gone. [40][41][42][43]

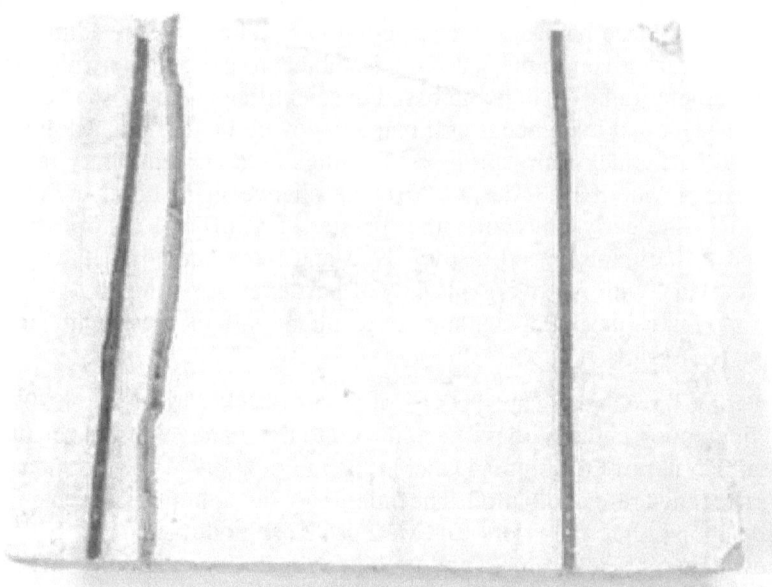

Figure 4 Black Rain Wall. A section of wall from a building far enough away from the blast to have remained standing but close enough to have sustained damage. The rain penetrated the damaged roof and ran down the walls leaving the tracks seen on the plaster. The section of wall is a display at the Oak Ridge Associated Universities Museum.[44] The picture above gives some idea of the "sump oil" nature of the black rain. Reproduced with permission of Oak Ridge Associated Universities Museum.

The need to expand the area designated as being affected by the Black Rain has been long standing and a number of scientists have advocated for it. The reason for this is not merely technical, it springs from a sense of justice and from human compassion toward those who have suffered, and who still suffer the debilitating effects of the A-bomb's Black Rain in Hiroshima. Telling the truth is mandatory.

While the people affected were out of range of the radiation pulse of mixed gamma and neutron rays, they were in the range of the rain created by the vast fires ignited by the nuclear blast. The smoke passed through the fireball and fell in black, thick, sticky rain drops.[26] [44]

Of Syndromes and Conditions

The story of Dr. Hida of Hiroshima is widely known. Dr. Hida has treated people affected by the atomic bombs since 1945. His view of the effects suffered differs markedly from that of the conventional view and description.[27] On July

12, 2012, *The Japan Times* published an article relating Dr. Hida's knowledge of the effects of the atomic bomb upon survivors – both survivors who had survived acute radiation syndrome and those who, like the Black Rain survivors, had not been subject to the bomb's radiation pulse. The latter group included people such as those who entered Hiroshima after the bomb had detonated, looking for loved ones. Dr. Hida told *The Japan Times* that many survivors were blighted by the condition known as Bura Bura disease in Japan:

> The illness haunted thousands of atomic-bomb survivors, including those who escaped the direct blast but inhaled, drank or ate radioactive substances, he says. Those who exhibited the symptoms felt too tired to work or even stand up, but doctors could not clearly establish they were ill. The patients lost trust in society as they were regarded by some as pretending to be sick or were just being lazy.[45]

Dr. Hida contends that Bura Bura disease derives from acute and chronic radiation exposure.

Is there any confirmation in the public arena which confirms Dr. Hida's observation and diagnosis of Bura Bura disease as an outcome of chronic radiation exposure, such as that caused by living in a chronically contaminated environment? Of the material aimed at the general public, that provided by the US Environmental Protection Agency (EPA) has a long-standing importance. The EPA states:

> In general, the amount and duration of radiation exposure affects the severity or type of health effect. There are two broad categories of health effects: stochastic and non-stochastic...Stochastic effects are associated with long-term, low-level (chronic) exposure to radiation. ("Stochastic" refers to the likelihood that something will happen.) Increased levels of exposure make these health effects more likely to occur, but do not influence the type or severity of the effect. Cancer is considered by most people the primary health effect from radiation exposure... Other stochastic effects also occur. Radiation can cause changes in DNA, the "blueprints" that ensure cell repair and replacement produces a perfect copy of the original cell. Changes in DNA are called mutations. Sometimes the body fails to repair these mutations or even creates mutations during repair. The mutations can be teratogenic or genetic. Teratogenic mutations are caused by exposure of the fetus in the uterus and affect only the individual who was exposed. Genetic mutations are passed on to offspring.[46]

This description is brief and for the general information of the public. It does not mention any symptoms remotely like those of Bura Bura disease as described by Dr. Hida. Further, there is no mention of chronic effects related to chronic (long-term) exposure. People are also confronted with a rarely used word – stochastic- and a basic translation of it.

Yes, dose is very important. Very often the initial dose measurements are totally inadequate and must be reconstructed years later. Usually these dose reconstructions undervalue the contribution of internalised radioactive dust.[47] The dual concepts and labels of "non-stochastic" and "stochastic" under the broad headings encompass the Western view of the health effects of radiation exposure. The headings themselves give no hint of the existence of a health syndrome caused by chronic exposure to environmental radiological contamination.

Although reticent to consider effects from chronic, "low-level" exposure, the West has long accepted the condition known as Acute Radiation Syndrome and the complex path to recovery survivors must often face, if they survive at all. The US Centres for Disease Control do give a clear and detailed description of Acute Radiation Syndrome. The description is aimed at doctors as a means and as an aid to diagnosis of the syndrome. It states in part:

> Acute Radiation Syndrome (ARS) (sometimes known as radiation toxicity or radiation sickness) is an acute illness caused by irradiation of the entire body (or most of the body) by a high dose of penetrating radiation in a very short period of time (usually a matter of minutes). The major cause of this syndrome is depletion of immature parenchymal stem cells in specific tissues. Examples of people who suffered from ARS are the survivors of the Hiroshima and Nagasaki atomic bombs, the fire fighters that first responded after the Chernobyl Nuclear Power Plant event in 1986, and some unintentional exposures to sterilization irradiators.

The required conditions for Acute Radiation Syndrome (ARS) are:

- **The radiation dose must be large** (i.e., greater than 0.7 Gray (Gy) or 70 rads).

- Mild symptoms may be observed with doses as low as 0.3 Gy or 30 rads.

- **The dose usually must be external** (i.e., the source of radiation is outside of the patient's body).

- Radioactive materials deposited inside the body have produced some ARS effects only in *extremely rare cases.*

- **The radiation must be penetrating** (i.e., able to reach the internal organs).

- *High energy X-rays, gamma rays, and neutrons are penetrating radiations.*

- **The entire body** (or a significant portion of it) must have received the dose.

- Most radiation injuries are local, frequently involving the hands, and these local injuries seldom cause classical signs of ARS.

- **The dose must have been delivered in a *short time*** (usually a matter of minutes).

- Fractionated doses are often used in radiation therapy. These are large total doses delivered in small daily amounts over a period of time. Fractionated doses are less effective at inducing ARS than a single dose of the same magnitude.[48]

It is my view that the definition of Acute Radiation Syndrome is not merely a medical definition. It is in fact a ***Medico-legal definition.*** While this syndrome captures the effects of high levels of exposure, it fails to encompass the long-term effects of chronic exposure to "low-level" radionuclides in the environment.

The Genesis of Official Amnesia and an Un-Studied Syndrome

Upon the surrender and occupation of Japan, the US established the Atomic Bomb Casualty Commission (ABCC). This entity was not a joint Japan-USA body, as is the current successor organization, the Radiation Effects Research Foundation. From its formation in 1946, the ABCC took orders only from American authorities.[49] In the days following the atomic bombings, Japanese doctors raised disagreements with members of the ABCC:

The Japanese doctors reminded the ABCC investigators that it had rained after the detonations. Dr. Yamazaki asked if the rain had been black and Dr. Odachi said he thought it was. Dr Sugihara remembered, "It was as though looking through a veil. One could gaze and clearly define the outline of the sun without feeling the glare". [50]

Concerns about black rain were ignored by American authorities.

The Occupation of Japan ended shortly after the signing of the Peace Treaty in San Francisco on April 28, 1951. This treaty was between Japan and 48 nations. The Treaty came into effect in 1952. The Soviet Union did not sign the treaty at that time.[51] The ABCC was an entity that was an organ of the Occupation of Japan. The medical literature relating to the bomb was heavily censored.[52] The exchange between the Japanese doctors and the Atomic Bomb Casualty Commission shows that the Americans were not so interested in the people who were affected by the Black Rain. The ABCC had to be reminded.

Is there any reference anywhere from a qualified source that verifies Doctor Hida's observations of a condition caused by radiation which inflicts a very long lasting, constant, debilitating fatigue and susceptibility to recurrent illness upon victims? Are there findings relating to a syndrome that is not the result of an acute one-time high dose exposure from an external source to the whole body? Is there a radiation-caused syndrome that causes multi-system disease states within the person and that occurs in response to chronic, but lower dose and dose rates, than that of ARS? Is there, in the records of any other country, documentation of Bura Bura disease by another name? When was it discovered? Who discovered it? What did the West do with this knowledge? Is there a treatment? *Official ignorance of health outcomes is the cause of the dichotomy between nuclear victims and nuclear authorities. I believe this official ignorance is willful and universal in the West.*

Mayak Disease

In 1971, a book entitled *Radiation Sickness in Man (Outlines)* was published in Moscow. It had been written by two Soviet doctors named A.K. Guskova and G.D. Baisogolov. It was one of those Soviet books that had been published in English and Russian.[53] In this book the Soviet authors describe the "new" radiation-related disease they identified, described and diagnosed in people living in the Soviet Union. They named this "new" disease "Chronic Radiation Syndrome".[54] The book was considered to be so important by the USA that the US Atomic Energy Commission republished it in 1973.[55]

This disclosure of Soviet radiation victims in this book predates the disclosures of Z. A. Medvedev regarding Soviet radiation victims in the Urals by some five years.[56] [57] [58] Yet, the conventional chronology regarding the Western discovery of the Urals nuclear disaster is rendered sadly lacking by the tendency to omit the 1971 Soviet publication of the work by Guskova and Baisogolov. This is an unfortunate omission since their 1971 work announces the discovery, diagnosis and treatment of a deterministic, chronic, multi-organ system disease syndrome caused by both internal and external sources of radiation, and which is induced not by acute exposure, but by chronic exposure to ionising radiation at doses less than that required to induce Acute Radiation Syndrome - and over much longer time spans.[53]

The Urals Research Centre for Radiation Medicine describes the radiation research of A. K. Guskova as follows:

> Angelina Konstantinovna Guskova, (born 1924), radiologist, MD, professor, winner of the Lenin Prize (1963), a member of the National Commission on Radiological Protection (since 1959), expert of the United Nations Scientific Committee on the Effects of Atomic Radiation (since 1967), corresponding member of the Academy of Medical Sciences (1986), Chief Scientific Officer/chief researcher of the Burnazyan Federal Medical Biophysical Center of Federal Medical Biological Agency of Russia, Honored Science Worker of the RSFSR

(1989), laureate of the Sievert Award for her contribution to radiation protection (2000). The main line of scientific research and main practical achievements of A.K. Guskova can be represented as follows: development of fundamental etiopathogenetic classification of radiation sickness in humans (together with G. D. Baysogolov), direct involvement into the treatment, evaluation of its effectiveness and formation of the basic principles of therapeutic and diagnostic measures in radiation accidents of various types, and participation in the system of preventive measures for the staff of Mayak PA that led to the health recovery in the vast majority of people (88 percent) out of thousands of exposed persons, participation in the work of the United Nations Scientific Committee on the Effects of Atomic Radiation (UNSCEAR) and the preparation of the Committee's reports concerning acute radiation effects, clinical radiation epidemiology, radiation effects on the nervous system, and involvement in the program on cardiovascular disease (radiation contribution to polyetiologic diseases).[59]

Guskova's work with Baisogolov indisputably documents the effects of chronic, "low-level" radiation exposure, despite languishing until re-discovered in 1994.

The United States Armed Forces Radiobiology Research Institute, Bethesda, MD in 1994 contracted the Urals Research Center for Radiation Medicine, Chelyabinsk, Russia to write the paper "Analysis of Chronic Radiation Sickness Cases in the Population of the Southern Urals". This work was undertaken under contract number DNAOO1-92-M-0658, issued by the then Defense Nuclear Agency of the US. This cooperation arose out of the nuclear emergency that was created by the Chernobyl nuclear disaster of 1986 and out of the collapse of the Soviet Union which followed shortly thereafter.[54]

A study of the nature of the Soviet "Chronic Radiation Syndrome" documented in this report and in the book by Guskova and Baisogolov follows in order to see if this disease syndrome in any way provides a second source for the disease identified in Black Rain survivors and post-bombing city entrants by Japanese doctors in 1945 and identified by those doctors as being "Bura Bura Disease".[27] Is the Soviet/Russian "Chronic Radiation Syndrome" really a "new" disease or has the West been deliberately blind to it since 1945?

The United States Discovers Chronic Radiation Syndrome.

The US-sponsored 1994 report, *Analysis of Chronic Radiation Sickness Cases in the Population of the Southern Urals*, edited by Mira M. Kossenko, compiles radiation exposure effects for workers at the Mayak plutonium production plant, drawing upon previous work published by Guskova and Baisogolov. The report indicates that Mayak workers were exposed chronically to radiation levels much higher than workers in the west were usually subjected to:

> ...**because of the long-term exposure to levels of ionizing radiation that were often orders of magnitude above those Western workers generally experienced, several individuals, according to the literature, reported, symptoms of sleep and appetite disturbances, difficulties with concentration and memory, irritability, and other "soft" clinical signs.** Complete blood counts, when taken, revealed pancytopenia. Symptoms would improve and counts would return to normal **only when the individual was removed from sources of radiation exposure.** A team of physicians headed by Dr. A. K. Guskova and Dr. G. D. Baysogolov coined the term "**chronic radiation sickness**" to describe these effects, which they felt to be due to the unusually high levels of radiation received and the length of exposure. Their work is cited in this report. In 1989, the veil of secrecy was lifted.
>
> According to its records, the International Atomic Energy Agency was notified of the accident at Kyshtym (near Mayak), where an underground tank with highly concentrated wastes exploded in 1957 and contaminated a large area. Scientists from **Branch 4 of the Institute of Biophysics, now the Urals Research Center for Radiation Medicine (URCRM),** were permitted to disclose the results of the dosimetric and clinical investigations conducted earlier.
>
> Dr. A. A. Akleyev, Dr. M. M. Kossenko, and Dr. M. O. Degteva visited the Armed Forces Radiobiology Research Institute (AFRRI) and presented some of their data. In June 1992, under the leadership of Professor V. N. Soyfer, a historic workshop was conducted at George Mason University in Fairfax. VA. This workshop, which was underwritten by AFRRI and the Department of Energy, brought together scientists and political figures from both the Russian Federation and the United States. As a result of those and subsequent discussions, AFRRI and URCRM collaborated in studying the effects of chronic radiation exposure on the Techa river village populations. It is hoped that the joint effort, which resulted in this report, will be the springboard for further research.[60]

The historic workshop of 1992 shows that the US was very interested to learn from and work with the Russian scientists. The reason why no historic workshop has been conducted between the US and Dr. Hida remains mysterious.

As explained above, *Analysis of Chronic Radiation Sickness Cases in the Population of the Southern Urals* draws upon work by Guskova and Baysogolov in defining chronic radiation sickness:

> Chronic radiation sickness is a complex, clearly outlined clinical syndrome occurring as a result of the long- term exposure of the organism to radiation, single or total doses of which regularly exceed the dose permissible for professional exposure.

This definition does not quantify the permissible irradiation doses for plant personnel. According to the radiation safety standards adopted in the USSR, these norms varied across time. Chronic radiation sickness was defined less in terms of quantified exposure levels and more in terms of bioaccumulation:

> Chronic radiation sickness is characterized by a certain dynamics of the clinical course directly related to radiation load formation, a combination of slow build-up of radiation affections and signs of compensatory processes and adaptive reactions. Individual symptoms and even clinical syndromes are not characteristic exclusively of radiation sickness but their very sequence may be considered as a characteristic feature allowing [us] to distinguish chronic radiation sickness as a separate clinical entity. As a rule, a correct, well-grounded diagnosis of chronic radiation sickness caused by general irradiation does not present a great difficulty and may be established at any therapeutic or prophylactic institution.....

Analysis of Chronic Radiation Sickness Cases in the Population of the Southern Urals distinguished three degrees of gravity in classifying chronic exposure. It was emphasized, however, that the distinction of degrees of gravity of CRS is to a certain degree a matter of convention:

> CRS of first degree of gravity (mild) is characterized by neuro-regulatory disorders in different organs and systems (cardiovascular system in particular), presence of unstable moderate leukopenia, and less frequently thrombocytopenia. The second degree (medium gravity) is associated with more pronounced regulatory disorders accompanied by the development of functional insufficiency of digestive glands and the cardiovascular and nervous systems, and signs of anatomic damage of radiosensitive tissues, hypoplastic status of hematopoiesis, changes in the myelin of the CNS conduction tracts, and disturbances of some metabolic processes: The third degree of gravity (severe) is characterized by destructive processes in the hematopoietic tissue, atrophic changes in the mucous membrane of the gastrointestinal tract, myocardial dystrophy, disseminated encephalomyelosis with a mild course,

and, in cases of weakened general immunity, infectious/septic complications.

CRS was applied in practice as early as 1950 by physicians serving personnel who were producing plutonium for weapons at Mayak.[61]

The 1994 Russian contractors give as an original source document for their definition of Chronic Radiation Syndrome as *"Guskova AK, Baysogolov GD (1971) Radiation Diseases in Humans.* **Moscow, page 387"[62].** As we have seen, the 1971 *"Radiation Sickness in Man (Outlines)"* published by Guskova and Baisogolov is a seminal source documenting chronic exposure effects. This book was re-published in the United States in 1973 by the US Atomic Energy Commission.[55] The record clearly shows that the disclosures of Guskova and Baisogolov, as published in English by the USSR in 1971, in regard to the existence of Chronic Radiation Syndrome have been known to nuclear authorities in the USA and probably the West in general, not since 1992, not since 1989, but since 1971. The "historic" disclosures in fact occurred 18 years earlier than claimed by the Armed Forces Radiobiology Research Institute in its 1994 publication. The question which arises out of this is: What were the nuclear authorities of the US doing with the information regarding Chronic Radiation Syndrome in the period since 1971? Why did the US wait until 1992 to have a "historic" workshop with representatives of the Russian Institute responsible for the identification, diagnosis and treatment of Chronic Radiation Syndrome in the USSR from 1950 on? If the Cold War was such an impediment – despite the myriad other scientific East–West collaborations which occurred in that period – why was the US Atomic Energy Commission given permission to republish the text in question by the Soviet Union? These questions remain unanswered.

Chronic Radiation Syndrome and Bura Bura Disease

A more recent Russian publication titled "Chronic Radiation Syndrome (CRS) in Residents of the Techa Riverside Villages" by A.V. Akleyev, published in May 2013, gives a detailed summary of the components of Chronic Radiation Syndrome:

Changes of hematopoiesis.

In the peripheral blood: leukopenia, neutropenia with left-shift in leukogram, thrombocytopenia, rarely–lymphocytopenia.

In the Bone Marrow: delay in the BM granulocyte maturation at the stage of myelocyte, decrease in the activity of megakaryocytes, increase of proliferative activity, and accelerated maturation of erythrokaryocytes, increased level of aberrant neutrophilic and erythroid cells at the stage of mitosis or at interphase.

Neurological disorders.

Vegetative dysfunction (arterial hypotonia, bradycardia, disturbed motor and secretory functions of the digestive organs, etc.).

...Asthenic syndrome (significant weakness, increased fatigability, sleep disorders, etc.).

...Microorganic disorders of the nervous system (persistent nystagmus, static ordynamicataxia, muscularhypotonia, anisoreflexia of the tendon and periosteal reflexes, pathological reflexes, etc.).

...CRS represents a systemic response of the body as a whole to the chronic total body exposure in man.

...At the initial stage, CRS can be defined as a "disregulatory" pathology which is formed on the basis of radiation-induced disorders in the regulatory systems of man (nervous, endocrine and immuno-hematopoietic). The changes in the cardio-vascular, digestive, reproductive, and other systems, are of a secondary functional nature, and are reversible....

Higher doses to regulatory and visceral organs induce irreversible organic alterations (vascular disorders, dystrophy, fibrosis, BMhypoplasia, etc.), and the course of CRS may assume an irreversible character.[63]

This report merely verifies early work published by *Guskova and Baysogolov GD* in 1971 by the USSR and in 1973 by the US.

I conclude that the West has known of the definition of Chronic Radiation Syndrome and of its repeated diagnosis and treatment in the USSR, then Russia, since 1971 at the latest. How useful would this knowledge have been had US authorities let those affected by the Nevada nuclear weapons tests, the Three Mile Island reactor meltdown and other incidents know that the AEC had possession of information regarding a "new" radiation-induced disease of a chronic nature? At issue is a disease syndrome which in fact many US Downwinders suffered from, complained about and sought justice for in US courts at that time. Victims typically lost legal contests because of a claimed "lack of evidence". Access to and formal recognition of the 1971 report would afford nuclear test veterans and victims the world over additional just cause to make their medical claims.

Today, Bura Bura disease (as first described in 1945) and Chronic Radiation Syndrome (as first described in the USSR in 1950) are recognised by the US military. When are the nuclear authorities planning on telling the people of Japan, of Nevada, of the Marshall Islands, of Maralinga, Australia – and on and on and on about a syndrome they have secretly studied for decades? Maybe they will disclose radiation realities only when Westinghouse, General Electric, Toshiba and Mitsubishi give them the green light to do so.

Dose is a medical term, a technical term and a legal term. Let us see what the nuclear authorities cook up for us as the minimum dose deemed permissible before a diagnosis of "Chronic Radiation Syndrome" can be made in the medico-legal jungle which is the modern nuclear world. "Dose" is clearly political. It is my observation that with each nuclear disaster there is an assurance of safety – on the basis of measured dose. Then, some time later – sometimes within years, sometimes within decades – there come disclosures that crucial dose measurements were omitted or wrong originally. Will this happen in Japan post-Fukushima. It is a safe bet that it will in my opinion.

There is a documented tendency in Japan to refuse to acknowledge the full risks posed by exposure to past and present radiation disasters. The Japanese government in the latter part of 2012 refused to grant an application to extend the area of Hiroshima deemed to be affected by the Black Rain resultant from the atomic bombing of that city in 1945 unjustly. The decision was made in spite of the evidence and in spite of the findings in favour of the extension by multiple Japanese courts. The Japanese government made its determination not in reference to the due and just consideration of the people affected. It made its decision, I submit, in reference to the continuing disaster unfolding in the close-in areas affected by the multiple failures of nuclear power plants located in Fukushima Prefecture. For it wanted the evacuation zones and the exclusion zone to be as small and as economic as possible, despite the harm this would likely cause by way of increased risk, to people who have no say in their own eventual fate.

Those who complained about arbitrary demarcations were labelled as "weak minded" and naturally "unhappy" people. Many lessons apparently come from the Soviet experience, including the abuse of psychiatry. Too many psychiatric authorities have denied the neuro-biological effects of Chronic Radiation Syndrome, by emphasizing instead victims' purportedly irrational fear and weak mindedness. The psychologized term "radiophobia" is used without any reference to the body of knowledge relating to the effect of chronic radiation exposures which has amassed in fact since 1950.[69] The true effects of chronic radiation exposure are denied by psychological trivialization.

Today the ICRP is deciding at what dose point the disease Chronic Radiation Syndrome may be diagnosed.[64][65] No nuclear authority has explained any of this information to the ordinary people of the world. The day of the Nuclear Elite must pass. Dose is political. If you don't want a compulsory dose, vote against the technology that produces it. The inability to do that is testament to the corrosive effect nuclear technology has always had upon democracy.

I cannot find any evidence to support the claim by the Urals Research Centre for Radiation Medicine that "88 percent" of thousands of exposed persons were cured of their radiogenic illnesses by that organisation's methods and treatments.[59] To the contrary, I find that Mayak is an ongoing nuclear disaster in which many ordinary people continue to suffer chronic radiation exposure, resultant disease, illness and premature death.[66] [67] [68]. I find that Dr. Hida and the other Japanese doctors of Hiroshima and Nagasaki have been correct since 1945. It took the USSR only until 1950 to find Bura Bura Disease. They have called it "Chronic Radiation Syndrome". The reports of Chronic Radiation Syndrome – Bura Bura Disease - by Japanese doctors and the recorded sufferings of Japanese civilians have been willfully ignored by Western governments since 1945. It is no surprise that the government of Japan instituted a new secrecy regime[70] prior to the monstrously premature return of populations to living spaces within the Fukushima exclusion zone.[71]

Omission and suppression of evidence of exposure effects can be found in the Japanese government's response to the Fukushima Daiichi nuclear disaster. The Japanese government continues denial of a disease syndrome known in Japan since 1945. The findings on chronic radiation exposure indicate that Japan's exclusion zone for the Fukushima Daiichi disaster was way too small from the beginning. As it was in 1945, so it is in 2014. *The Japan Times* reported on the zone:

> On March 17, 2011, the U.S. Embassy advised Americans to stay outside an 80-km radius of the stricken plant, while the Japanese government's evacuation order was for people within a 20-km radius... [The then Chairman of the NRC, Gregory] Jaczko stated "Because of the compelling need in a crisis to act and to make decisions, we proceeded to make predictions . . . what we found from these analyses was that the radiation releases would . . . potentially extend out to distances of 20, 30, 40 and 50 miles."[72]

The idea that people should ordinarily live and work within a 50 mile radius of a ruined nuclear power plant is seen for what it is: a life of imposed risk. The only way to avoid the imposition of risk by the nuclear industry is to recognize the hazards posed by the installations

The reality is that the hazards of Fukushima Daiichi are not specific, they are generic. The neat fix of locating nuclear power plants "away from major cities", as suggested by Lapp, has never been a solution. In the world of 2014, where is there a place remote enough to safely site a nuclear power plant?

During my service in the Australian Army I was instructed that atomic bombs are "multiple effects weapons."

One of the effects of atomic bombs is the generation of nuclear fallout.

In the wake of every nuclear reactor accident, nuclear industry has claimed that those who confuse atomic bombs with nuclear reactors are in error.

Nuclear reactors are not atomic bombs. However, both atomic bombs and failing, crippled, overheated and exploding nuclear reactors release uncontained radioactive material into land, water and air. In both cases this release of nuclear exhaust is technically termed "nuclear fallout."

Further, the atomic bomb which destroyed the city of Hiroshima in August 1945 contained 64 kg (141 lb) of uranium.[73]

The reactor cores of the failed reactors located at Fukushima Daiichi contained many tons of both fission fuel and fission products.[74]

On the basis of the past record of openness displayed by nuclear authorities, the people of the world will probably never be told the precise amount of nuclear material that entered the biosphere as fallout following the March 2011 nuclear disaster. It is my opinion that the experience of ordinary people will form a witness to the truth which clashes severely with the official accounts of the nuclear hierarchy. Just as occurred in the Black Rain suburbs of Hiroshima – and just as it did in every country where communities of nuclear "Downwinders" live.

Social isolation is a symptom of nuclear fission technology. An old technology, one patented in 1934 by Leo Szilard in his two patents. (440,023 Provisional Specification No. 7840 1934 Improvements in or relating to the Transmutation of Chemical Elements Accepted Dec 12 1935; 630,726 Provisional Specification No. 19157 1934 Accepted March 30, 1936 [but withheld from publication under Section 30 of the Patent and Designs Acts 1907 to 1932.] Date of publication 1949.) The industrial scale creation of nuclear poisons released into the biosphere is the result of a technology which was past its 'use by' date in the 1930s.

I am very grateful to the Hiroshima Peace Memorial Museum and to the Oak Ridge Associated Universities Museum for their very kind permission to use the photographic images they have provided.

References

[1] BBC material rebroadcast by SBS TV Australia, 15 March 2011.

[2] Aljazeera, "Radiation warning after plant blast", 15 March 2011,
http://www.aljazeera.com/news/asia-pacific/2011/03/
201131584630423499.html)

[3] "Emergency Core Cooling", W K Ergen; U.S. Atomic Energy Commission. Advisory Task Force on Power Reactor Emergency Cooling. Published by the US Atomic Energy Commission, Division of Technical Information, Oak Ridge, USA, in 1967, pp 166 -168

[4] Constructing Fukushima Diiachi, YouTube: 福島原子力発電所建設記録 黎明 調査篇 2
http://www.youtube.com/watch?feature=player_embedded&v=7g6UFAs z_LM , uploded by Hiroshi Taguchi.

[5] Wikipedia, citing "Japan: Nuclear Power Reactors". Power Reactor Information System – PRIS. IAEA. Exact date of start of construction given as 25 July 1967. http://en.wikipedia.org/wiki/Fukushima_Daiichi_Nuclear_Power_Plant

[6] ABC Radio, Australia. "AM", Mark Willacy, Fukushima radiation kills fishing industry Posted Mon 25 Jun 2012, 3:16pm AEST. http://www.abc.net.au/news/2012-06-25/fukushima-radiation-kills-fishing-industry/4091320

[7] Fukushima plant spilling contaminated water into the sea 'for years', AM ABC Australia By North Asia correspondent Mark Willacy, ABC Radio, Australia, 12 August 2013. http://www.abc.net.au/news/2013-08-12/fukushima-plant-workers-raise-safety-concerns/4879960

[8] "Fukushima Daiichi ANS Committee Report section II.B.2. Fukushima Daiichi Unit 2. http://fukushima.ans.org/report/accident-analysis

[9] "Inside Japan's Nuclear Meltdown", Dan Edge, BBC, Quicksilver productions, PBS, 28 February 2012. http://www.pbs.org/wgbh/pages/frontline/japans-nuclear-meltdown/ video, transcript: http://www.pbs.org/wgbh/pages/frontline/health-science-technology/japans-nuclear-meltdown/transcript-4/

[10] "The menace of atomic energy", Chapter: "The ECCS Controversy", Abbott, J., Nader, R., Norton, New York, ISBN 0393087735

[11] "Inherently Safe Reactors", I Spiewak, and A M Weinberg, Annual Review of Energy Vol. 10: 431-462 (Volume publication date November 1985). DOI: 10.1146/annurev.eg.10.110185.002243Institute for Energy Analysis, Oak Ridge Associated Universities, http://www.annualreviews.org/doi/abs/10.1146/annurev.eg.10.110185.002243?journalCode=energy

[12] "Thoughts on Nuclear Plumbing", Lapp, R., *The New York Times*, 12 December 1971.

[13] "Flow Accumulation Near Fukushima Daiichi PP", Pacific Disaster Centre, http://reliefweb.int/sites/reliefweb.int/files/resources/3E04DD1DD13989BC8525785D004CCD7A-map.pdf

[14] "The Geology of Fukushima, by Pierre Fetet, Translation from French: Robert Ash, http://ddata.over-blog.com/4/37/62/00/The-Geology-of-Fukushima.pdf

[15] "TEPCO took no action about radioactive water leak for 2 years", *The Ashai Shimbun*, 1 August 2013. http://ajw.asahi.com/article/0311disaster/fukushima/AJ201308010053

[16] "Fukushima nuclear power plant poses new threat to Pacific Ocean", Lewis, L., *The Australian*, 6 August 2013. http://www.theaustralian.com.au/news/world/fukushima-nuclear-power-plant-poses-new-threat-to-pacific-ocean/story-fnb64oi6-1226692014141

[17] "Fukushima's Radioactive Water Leak: What You Should Know", Kliger, P., *National Geographic*, 7 August 2013.

[18] Aljazeera, "Radiation warning after plant blast", 15 March 2011, http://www.aljazeera.com/news/asia-pacific/2011/03/201131584630423499.html

[19] Sci Rep. 2013;3:1742. doi: 10.1038/srep01742. Overview of active caesium contamination of freshwater fish in Fukushima and Eastern Japan. Mizuno T., Kubo H. http://www.ncbi.nlm.nih.gov/pubmed/23625055)

[20] Atmospheric dispersion of radionuclides from the Fukushima-Daichii nuclear power plant, CEREA joint laboratory, École des Ponts ParisTech and EdF R&D Victor Winiarek, Marc Bocquet, Yelva Roustan, Camille Birman, Pierre Tran, Map of ground deposition of Caesium-137 for the Fukushima-Daiichi accident (updated 20 June 2013). http://cerea.enpc.fr/en/fukushima.html

[21] ABC TV, Australia, "The Drum" 16 March 2011. Dr. Jim Green quotes the statement from Adelaide's oldest University issued by one of its scientists, Barry Brooks, during the initial days of the emergency at Fukushima Diiachi. http://www.abc.net.au/unleashed/45210.html .

[22] Rick Wallace, "The Weekend Australian Magazine", March 10-11, 2012, p.22.)

[23] Asahi Shimbun newspaper, series of articles describing the nuclear disaster, including the SPEEDI Deception. The series is called "The Prometheus Trap" and is available here: http://ajw.asahi.com/tag/PROMETHEUS%20TRAP

[24] Greenpeace, "Call to widen evacuation area around Fukushima Blogpost by Brian Fitzgerald – March 27, 2011 at 9:38, http://www.greenpeace.org/usa/en/news-and-blogs/campaign-blog/call-to-widen-evacuation-area-around-fukushim/blog/33981.

[25] IAEA, Fukushima Nuclear Accident Update log, updates of 30 March 2011, "First assessment indicates that one of the IAEA operational criteria for evacuation is exceeded in Iitate village. We advised the counterpart to carefully assess the situation. They indicated that they are already assessing." IAEA website, http://www.iaea.org/newscenter/news/2011/fukushima300311.html)

[26] Masuda Yoshinobu, From "Black Rain" to "Fukushima": The Urgency of Internal Exposure Studies, *The Asia-Pacific Journal*, Vol 10, Issue 39, No. 3, September 24, 2012 http://japanfocus.org/-Masuda-Yoshinobu/3836#sthash.fufrROB1.dpuf

[27] "Account of a Medical Doctor Who Had to Face Innumerable Deaths of Victims from the Exposure to A-bomb Radiation", Shuntaro Hida http://afsc.org/sites/afsc.civicactions.net/files/documents/Shuntaro%20Hida,%20Japan.pdf

[28] Interview with Dr. Nello Pace, Ph.D. Setting: August 16, 1994; Berkeley, CA Interviewer: Anna Berge (Lawrence Berkeley Laboratory Archives and Records Office) DOEEH-0476 June 1995.

[29] Personal correspondence with the National Secretary of the Australian Atomic Ex-Servicemen's Association, Mr. Terry Toon. This association accepts members who have been exposed to radiation due to military occupation. I am a volunteer research officer with this organisation. The Association includes among its members that small remanent of members who saw service as Australian Occupation troops in Hiroshima and Kure from 1946 until 1952. The association also includes, of course, survivors of the British nuclear weapons tests in Australia.

[30] Australian Repatriation Commission finding reproduced in "Atomic Fallout", Vol1. No.10, June 1990. The organ of the Atomic ExServicemen's Association. Inc. Canberra. Australia.

[31] "Atomic Fallout", Vol. 3 No. 2 June/July 2000. The organ of the Australian Atomic ExServicemen's Association. Inc. Canberra. Australia.

[32] "The Aftermath of Hiroshima and Nagasaki: The Emergence of the Cold War Radiation Research Bureaucracy", Advisory Committee on Human Radiation Experiments, Presidential Committee, 1994. Ref docs: Robert S. Stone, M.D., to Lieutenant Colonel H. L. Friedell, U.S. Engineer Corps, Manhattan District, 9 August 1945 ("As you and many others are aware, a great many of the people . .") (ACHRE No. DOE-121494-D-1). This reference can now be found in the ACHRE Final Report, US Congress, page 35. This is available for online viewing and download at http://archive.org/stream/ advisorycommitte00unit/ advisorycommitte00unit_djvu.txt.

[33] The Franck Report, full text: http://www.dannen.com/decision/franck.html

[34] Tomomitsu Miyazaki, Senior Staff Writer, Hiroshima Peace Media Centre. The article is available at: http://www.hiroshimapeacemedia.jp/mediacenter/ article.php?story=20120524110701783_en

[35] "Government review of "black rain" area in Hiroshima bombing now in final stage", Hiroshima Peace Media Centre, Kohei Okada and Michiko Tanaka, Hiroshima Peace Media Centre, April 20th 2012. Map. http://www.hiroshimapeacemedia.jp/mediacenter/article.php?story=20120416 140310746_en

[36] "Citizens' Nuclear Information Centre, Japan: "Certification of Sufferers of Atomic Bomb-Related Diseases". In particular, see note 1 at bottom of page. http://www.cnic.jp/english/newsletter/nit131/nit131articles/abombdisease.html .

[37] Radiation survey activities in the early stages after the atomic bombing in Hiroshima, Tetsuji Imanaka, Research Reactor Institute, Kyoto University, Kumatori-cho, Sennan-gun, Osaka, 590-0494 Japan.

[38] "Historical sketch of the scientific field survey of Hiroshima several days after the bombing" by Prof Sakae Shimizu, 1982, published by the Bulletin of the Institute for Chemical Research, Kyoto University, Volume 60, (2), 1982.

[39] "Sacred ground: a portrait of Hiroshima in the light of the August 1945 radiological surveys", by Paul J. Langley. National Library of Australia, Dewey Number: 363.17990952.

[40] Hiroshima Peace Memorial Museum virtual museum display at: http://www.pcf.city.hiroshima.jp/virtual/VirtualMuseum_e/exhibit_e/exh0307_e/exh03076_e.html

[41] May 1973, the Japanese Ministry of Foreign Affairs held a ceremony for the return of Hiroshima and Nagasaki A-bomb records confiscated by the U.S. during the Occupation (1945-52). Atomic Bomb Museum – "The Pursuit of Peace". http://www.atomicbombmuseum.org/5_timetable.shtml

[42] "Investigation of the Fate of U-235 from the Hiroshima A Bomb" Yoko Fujikawa, Kiyoshi Shizuma and Satoru Endo, Research Reactor Institute, Osaka, 2003. This is a study of samples of returned 1945 Hiroshima soil originally collected on 9 August 1945 by Yoshio Nishina et al., Institute of Physical and Chemical Research, and confiscated in September 1945 by US Occupation Forces. http://irpa11.irpa.net/pdfs/7c4.pdf

[43] "The Repatriation of Atomic Bomb Victim Body Parts to Japan, Natural Objects and Diplomacy", Lindee, M.S., Osiris, second series, Volume 13, 1998.

[44] "Black Rain" text provided by Oak Ridge National Laboratory courtesy of David Fields, Oak Ridge Associated Universities, Museum. http://www.orau.org/ptp/collection/hiroshimatrinity/blackrain.htm

[45] "A-bomb doctor warns of further Fukushima woes", Megumi Iizuka, Kyodo, *The Japan Times*, July 12, 2012.

[46] "Radiation Protection", Health Effects, EPA, United States. http://www.epa.gov/radiation/understand/health_effects.html

[47] Major Alan Batchelor, adviser to Australia's Nuclear Veterans. Document: Submission to the Governor General of Australia, 29 March 2011. Available at: http://nuclearhistory.wordpress.com/2011/04/10/letter-to-governor-general-from-major-alan-batchelor/

[48] "Acute Radiation Syndrome: A Fact Sheet for Clinicians" , United States, Centres for Disease Control. http://www.bt.cdc.gov/radiation/arsphysicianfactsheet.asp

[49] "The Atomic Bomb Casualty Commission in Retrospect", F.W. Putnam, Proceedings of the National Academy of Sciences of the United States of America, vol. 95 no. 10, May 12, 1998.

[50] Roff, Sue, "Hotspots: The Legacy of Hiroshima and Nagasaki", Cassell, 1995, pp 111.

[51] Treaty of Peace with Japan, Taiwan Documents Project, http://www.taiwandocuments.org/sanfrancisco01.htm

[52] Braw, M., The Atomic Bomb Suppressed, American Censorship in Occupied Japan, Sharpe, M.E., Armonk, New York, 1991.

[53] Radiation Sickness in Man (Outlines), A.K. Guskova and G.D. Baisogolov, Moscow, Meditisina, 1971. World Cat, http://www.worldcat.org/title/radiation-sickness-in-man-outlines-by-a-k-guskova-and-g-d-baysogolov/oclc/14432460?ht=edition&referer=di

[54] "Analysis of Chronic Radiation Sickness Cases in the Population of the Southern Urals", Defense Technical Information Centre Accession Number: Accession Number : ADA286238, Descriptive Note : Contract rept. Corporate Author : ARMED FORCES RADIOBIOLOGY RESEARCH INST BETHESDA MD Personal Author(s) : Kossenko, M. M. ; Akleyev, A. A. ; Degteva, M. O. ; Kozheurov, V. P. ; Degtyaryova, R. G. Full Text : http://www.dtic.mil/get-tr-doc/pdf?AD=ADA286238 Report Date : AUG 1994 Pagination or Media Count: 94, dated August 1994, United States of America. Download link above located at US Defense Technical Information Centre, United States of America.

[55] "Radiation Sickness in Man (Outlines) by A.K. Guskova and Grigorii Davidovich Baisogolov, [Oak Ridge, Tenn., U.S. Atomic Energy Commission, Technical Information Center; Available from National Technical Information Service, U.S. Dept. of Commerce, Springfield, Va., 1973] Series United States Atomic Energy Commission. Technical Information Centre. Translation Series, AEC-tr-7401, World Cat, http://www.worldcat.org/title/radiation-sickness-in-man-outlines/oclc/756952&referer=brief_results

[56] "Two Decades of Soviet Dissidence", New Scientist, vol 72, no. 1025 (1976), pp 264- 267

[57] "Facts behind the Soviet Nuclear Disaster", New Scientist, vol 74, no 1058, (1977), pp 761 – 764.

[58] Medvedev, Z.A., "Nuclear Disaster in the Urals", Angus and Robertson Publishers, ISBN 0 207 95896 3, 1979.

[59] Urals Research Centre For Radiation Medicine, Federal Medical-Biological Agency of Russia. Federal State Research Institution. http://www.urcrm.ru/en/collaboration/professors.html

[60] Analysis of Chronic Radiation Sickness Cases in the Population of the Southern Urals, pp. i, ii.

[61] "Analysis of Chronic Radiation Sickness Cases in the Population of the Southern Urals", pp. 21, 22.

[62] "Analysis of Chronic Radiation Sickness Cases in the Population of the Southern Urals", reference 20, page 86.

[63] Federal Medical-Biological Agency (Russia) Urals Research Centre for Radiation Medicine "Chronic Radiation Syndrome (CRS) in residents of the Techa riverside villages", A.V. Akleyev, ConRad, 13-16 May 2013, Munich. http://media.bsbb.de/Conrad/AKLEYEV.pdf

[64] Wikipedia at http://en.wikipedia.org/wiki/Chronic_radiation_syndrome

[65] "Early and late effects of radiation in normal tissues and organs: threshold doses for tissue reactions and other non-cancer effects of radiation in a radiation protection context" The link for it, as given by Wikipedia, is here: http://www.icrp.org /docs/Tissue%20Reactions%20Report%20Draft%20for%20Consultation.pdf

[66] Greenpeace at: http://www.greenpeace.org/international/en/news/features/mayak-nuclear-disaster280907/

[67] The Greenpeace document, "Mayak: a 50 year tragedy" is available here: http://www.greenpeace.org/international/Global/international/planet-2/report/2007/9/mayak-a-50-year-tragedy.pdf

[68] In April 2012 Aljazeera journalists visited the Mayak affected area. Their report is located at : http://blogs.aljazeera.com/blog/europe/living-nuclear-hell

[69] The Russian Radiation Legacy: Its Integrated Impact and Lessons Environmental Health Perspectives 105, Supplement 6, December 1997 The Russian Radiation Legacy: Its Integrated Impact and Lessons, Marvin Goldman, Professor of Radiobiology Emeritus, University of California, Davis, California, http://users.physics.harvard.edu/~wilson/freshman_seminar/Radiation/publicat ions/The%20Russian%20Radiation%20Legacy%20Its%20Integrated%20Impa ct%20and%20Lessons.txt

[70] Japan's State Secrets Law: Hailed By U.S., Denounced By Japanese, by Lucy Craft, December 31, 2013 6:53 PM ET, NPR Parallels, http://www.npr.org/blogs/parallels/2013/12/31/258655342/japans-state-secrets-law-hailed-by-u-s-denounced-by-japanese

[71] "Fukushima residents cleared to return home amid ongoing contamination fears", by Euan McKirdy, CNN, April 1, 2014 -- Updated 1012 GMT.

[72] "Jaczko recalls chaos of Fukushima early days", Kazuaki Nagata, The Japan Times, March 12, 2014. http://www.japantimes.co.jp/news/2014/03/12/national/jaczko-recalls-chaos-of-fukushima-early-days/#.U-yciuN_t8E

[73] However, the *Manhattan Project Heritage Preservation Association Inc.* gives the amount of uranium fuel as 140 pounds. Please see: http://www.mphpa.org/classic/HISTORY/little_boy.htm

[74] Katherine Harmon. How Much Nuclear Fuel Does the Fukushima Diiachi Facility Hold? *Scientific American,* 17 March, 2011, *http://www.scientificamerican.com/article/nuclear-fuel-fukushima/*. The article

states that the pools at each reactor are thought to have contained the following amounts of spent fuel, according to *The Mainichi Daily News*: "Reactor No. 1: 50 tons of nuclear fuel, Reactor No. 2: 81 tons, Reactor No. 3: 88 tons, Reactor No. 4: 135 tons, Reactor No. 5: 142 tons, Reactor No. 6: 151 tons. Also, a separate ground-level fuel pool contains 1,097 tons of fuel; and some 70 tons of nuclear materials are kept on the grounds in dry storage. The reactor cores themselves contain less than 100 tons of fuel."

Author's note: The following sources contain more detailed information regarding the radiological surveys of Hiroshima conducted by Japanese scientists:

1. "Historical Sketch of the Scientific Field Survey in Hiroshima Several Days after the Bombing" by Prof SakaeShimizu, 1982, published in Bulletin of the Institute for Chemical Research, Kyoto University, Volume 60, (2), 1982.

2. "Let's Look at the Special Exhibit, Academic Survey Team Records, A bomb damage survey activities". Online display, Hiroshima Peace Memorial Museum at:
http://www.pcf.city.hiroshima.jp/virtual/VirtualMuseum_e/exhibit_e/exh0702_e/exh070207_e.html#01

3. "The 1945 Radiation Surveys of Hiroshima Conducted by Japanese Scientists", Langley, P. Download PDF at:
https://www.academia.edu/7501572/The_1945_Radiation_Surveys_of_Hiroshima_Conducted_by_Japanese_Scientists

ENERGY TRANSITIONS

50 Reasons to Fear the Worst from Fukushima

Harvey Wasserman

Fukushima's[1] missing melted cores and radioactive gushers continue to fester in secret.

Japan's harsh dictatorial censorship has been matched by a global corporate media blackout[2] aimed—successfully—at keeping Fukushima out of the public eye.

But that doesn't keep the actual radiation out of our ecosystem, our markets ... or our bodies.

Speculation on the ultimate impact ranges from the utterly harmless to the intensely apocalyptic[3].

But the basic reality is simple: for seven decades, government Bomb factories and privately-owned reactors have spewed massive quantities of unmonitored radiation into the biosphere.

The impacts of these emissions on human and ecological health are unknown primarily because the nuclear industry has resolutely refused to study them.

Indeed, the official presumption has always been that showing proof of damage from nuclear Bomb tests and commercial reactors falls to the victims, not the perpetrators.

And that in any case, the industry will be held virtually harmless.

This "see no evil, pay no damages" mindset dates from the Bombing of Hiroshima to Fukushima to the disaster coming next ... which could be happening as you read this.

Here are 50 preliminary reasons why this radioactive legacy demands we prepare for the worst for our oceans, our planet, our economy ... ourselves.

1. At Hiroshima and Nagasaki (1945), the U.S. military initially denied[4] that there was any radioactive fallout, or that it could do any damage. Despite an absence of meaningful data, the victims (including a group of U.S. prisoners of war) and their supporters were officially "discredited" and scorned.

2. Likewise, when Nobel-winners Linus Pauling and Andre Sakharov correctly warned[5] of a massive global death toll from atmospheric Bomb testing, they were dismissed with official contempt ... until they won in the court of public opinion.

3. During and after the Bomb Tests (1946-63), downwinders in the South Pacific and American west, along with thousands of U.S. "atomic vets," were told their

radiation-induced health problems were imaginary[6] ... until they proved utterly irrefutable.

4. When British Dr. Alice Stewart proved (1956) that even tiny x-ray doses[7] to pregnant mothers could double childhood leukemia rates, she was assaulted with 30 years of heavily funded abuse from the nuclear and medical establishments.

5. But Stewart's findings proved tragically accurate, and helped set in stone the medical health physics consensus that there is no "safe dose" of radiation ... and that pregnant women should not be x-rayed[8], or exposed to equivalent radiation.

6. More than 400 commercial power reactors have been injected into our ecosphere with no meaningful data to measure their potential health and environmental impacts, and no systematic global data base has been established or maintained.

7. Acceptable dose" standards for commercial reactors were conjured from faulty A-Bomb studies[9] begun five years after Hiroshima, and at Fukushima and elsewhere have been continually made more lax to save the industry money.

8. Bomb/reactor fallout delivers alpha and beta particle emitters[10] that enter the body and do long-term damage, but which industry backers often wrongly equate with less lethal external gamma/x-ray doses from flying in airplanes or living in Denver.

9. By refusing to compile long-term emission assessments, the industry systematically hides health impacts at Three Mile Island (TMI), Chernobyl, Fukushima, etc., forcing victims to rely on isolated independent studies which it automatically deems "discredited."

10. Human health damage has been amply suffered in radium watch dial painting, Bomb production, uranium mining/milling/enrichment, waste management and other radioactive work, despite decades of relentless industry denial.

11. When Dr. Ernest Sternglass, who had worked with Albert Einstein, warned that reactor emissions were harming people[11], thousands of copies of his *Low-Level Radiation* (1971) mysteriously disappeared from their primary warehouse.

12. When the Atomic Energy Commission's (AEC) Chief Medical Officer, Dr. John Gofman, urged that reactor dose levels be lowered by 90 percent[12], he was forced out of the AEC and publicly attacked, despite his status a founder of the industry.

13. A member of the Manhattan Project, and a medical doctor responsible for pioneer research into LDL cholesterol, Gofman later called the reactor industry an instrument of "premeditated mass murder."[13]

14. Stack monitors and other monitoring devices failed at Three Mile Island (1979) making it impossible to know how much radiation escaped, where it went or who it impacted and how.

15. But some 2,400 TMI downwind victims and their families were denied a class action jury trial by a federal judge who said "not enough radiation" was released to harm them, though she could not say how much that was or where it went.

16. During TMI's meltdown, industry advertising equated the fallout with a single chest x-ray to everyone downwind, ignoring the fact that such doses could double leukemia rates among children born to involuntarily irradiated mothers.

17. Widespread death and damage downwind from TMI have been confirmed[14] by Dr. Stephen Wing, Jane Lee and Mary Osbourne, Sister Rosalie Bertell, Dr. Sternglass, Jay Gould, Joe Mangano and others, along with hundreds of anecdotal reports.

18. Radioactive harm to farm and wild animals downwind from TMI has been confirmed by the Baltimore News-American and Pennsylvania Department of Agriculture.

19. TMI's owner quietly paid out at least $15 million[15] in damages in exchange for gag orders from the affected families, including at least one case involving a child born with Down's Syndrome.

20. Chernobyl's explosion became public knowledge only when massive emissions came down on a Swedish reactor hundreds of miles away, meaning that—as at TMI and Fukushima—no one knows precisely how much escaped or where it went.

21. Fukushima's on-going fallout[16] is already far in excess of that from Chernobyl, which was far in excess of that from Three Mile Island.

22. Soon after Chernobyl blew up (1986), Dr. Gofman predicted its fallout[17] would kill at least 400,000 people worldwide.

23. Three Russian scientists who compiled more than 5,000 studies[18] concluded in 2005 that Chernobyl had already killed nearly a million people worldwide.

24. Children born in downwind Ukraine and Belarus still suffer a massive toll of mutation and illness, as confirmed by a wide range of governmental, scientific and humanitarian organizations.

25. Key low-ball Chernobyl death estimates come from the World Health Organization, whose numbers are overseen by International Atomic Energy Agency, a United Nations organization chartered to promote the nuclear industry.

26. After 28 years, the reactor industry has still not succeeded in installing a final sarcophagus over the exploded Chernobyl Unit 4[19], though billions of dollars have been invested.

27. When Fukushima Units 1-4 began to explode, President Obama assured us all the fallout would not come here, and would harm no one, despite having no evidence for either assertion.

28. Since President Obama did that, the U.S. has established no integrated system to monitor Fukushima's fallout,[20] nor an epidemiological data base to track its health impacts … but it did stop checking radiation levels in Pacific seafood.

29. Early reports of thyroid abnormalities[21] among children downwind from Fukushima, and in North America are denied by industry backers who again say "not enough radiation" was emitted though they don't know how much that might be.

30. Devastating health impacts reported by sailors stationed aboard the USS Ronald Reagan near Fukushima are being denied by the industry and Navy[22], who say radiation doses were too small to do harm, but have no idea what they were.

31. While in a snowstorm offshore as Fukushima melted, sailors reported a warm cloud passing over the Reagan[23] that brought a "metallic taste" like that described by TMI downwinders and the airmen who dropped the Bomb on Hiroshima.

32. Though it denies the sailors on the Reagan were exposed to enough Fukushima radiation to harm them, Japan (like South Korea and Guam) denied the ship port access because it was too radioactive (it's now docked in San Diego).

33. The Reagan sailors are barred from suing the Navy, but have filed a class action against Tokyo Electric Power (TEPCO)[24], which has joined the owners at TMI, the Bomb factories, uranium mines, etc., in denying all responsibility.

34. A U.S. military "lessons learned" report from Fukushima's Operation Tomodachi clean-up campaign notes that "decontamination of aircraft and personnel without alarming the general population created new challenges."

35. The report questioned the clean-up because "a true decontamination operations standard for 'clearance' was not set," thereby risking "the potential spread of radiological contamination to military personnel and the local populace."

36. Nonetheless, it reported[25] that during the clean-up, "the use of duct tape and baby wipes was effective in the removal of radioactive particles."

37. In league with organized crime, TEPCO is pursuing its own clean-up activities by recruiting impoverished homeless[26] and elderly citizens for "hot" on-site labor, with the quality of their work and the nature of their exposures now a state secret.

38. At least 300 tons of radioactive water continues to pour into the ocean[27] at Fukushima every day, according to official estimates made prior to such data having been made a state secret.

39. To the extent they can be known, the quantities and make-up of radiation pouring out of Fukushima are also now a state secret[28], with independent measurement or public speculation punishable by up to ten years in prison.

40. Likewise, "There is no systematic testing in the U.S.[29] of air, food and water for radiation," according to University of California (Berkeley) nuclear engineering Professor Eric Norman.

41. Many radioactive isotopes[30] tend to concentrate as they pour into the air and water, so deadly clumps of Fukushima's radiation may migrate throughout the oceans for centuries to come before diffusing, which even then may not render it harmless.

42. Radiation's real world impact becomes even harder to measure in an increasingly polluted biosphere, where interaction with existing toxins[31] creates a synergy likely to exponentially accelerate the damage being done to all living things.

43. Reported devastation among starfish, sardines, salmon, sea lions, orcas and other ocean animals cannot be definitively denied without a credible data base[32] of previous experimentation and monitoring, which does not exist and is not being established.

44. The fact that "tiny" doses of x-ray can harm human embryos portends that any unnatural introduction of lethal radioactive isotopes into the biosphere, however "diffuse," can affect our intertwined global ecology in ways we don't now understand.

45. The impact of allegedly "minuscule" doses spreading from Fukushima will, over time, affect the minuscule eggs of creatures ranging from sardines to starfish to sea lions, with their lethal impact enhanced by the other pollutants already in the sea.

46. Dose comparisons to bananas and other natural sources are absurd and misleading as the myriad isotopes from reactor fallout will impose very different biological impacts[33] for centuries to come in a wide range of ecological settings.

47. No current dismissal of general human and ecological impacts—"apocalyptic" or otherwise—can account over time for the very long half-lives of radioactive isotopes Fukushima is now pouring into the biosphere.

48. As Fukushima's impacts spread through the centuries, the one certainty is that no matter what evidence materializes, the nuclear industry will never admit to doing any damage, and will never be forced to pay for it (see upcoming sequel).

49. Hyman Rickover, father of the nuclear navy, warned that it is a form of suicide to raise radiation levels within Earth's vital envelope, and that if he could, he would "sink" all the reactors he helped develop.

50. "Now when we go back to using nuclear power," he said in 1982, "I think the human race is going to wreck itself[34], and it is important that we get control of this horrible force and try to eliminate it."

As Fukushima deteriorates behind an iron curtain of secrecy and deceit, we desperately need to know what it's doing to us and our planet.

It's tempting to say the truth lies somewhere between the industry's lies and the rising fear of a tangible apocalypse.

In fact, the answers lie beyond.

Defined by seven decades of deceit, denial and a see-no-evil dearth of meaningful scientific study, the glib corporate assurances that this latest reactor disaster won't hurt us fade to absurdity.

Fukushima pours massive, unmeasured quantities of lethal radiation into our fragile ecosphere every day, and will do so for decades to come.

Five power reactors have now exploded on this planet and there are more than 400 others still operating.

What threatens us most is the inevitable next disaster … along with the one after that … and then the one after that …

Pre-wrapped in denial, protected by corporate privilege, they are the ultimate engines of global terror.

See the article originally posted at:
http://ecowatch.com/2014/02/02/50-reasons-fear-fukushima/
(See more about the author at http://ecowatch.com/author/hwasserman/)

References

1. See more articles on the Fukushima nuclear power plant disaster at http://ecowatch.com/?s=fukushima

2. Ralph Nader, The Fukushima Secrecy Syndrome – From Japan to America http://nader.org/2014/01/24/fukushima-secrecy-syndrome-japan-america/

3. Harvey Wassermann, Japan's New 'Fukushima Fascism', http://ecowatch.com/2013/12/11/japans-new-fukushima-fascism/

4. Greg Mitchell, The Great Hiroshima Cover-up – And the Greatest Movie Never Made, http://japanfocus.org/-Greg-Mitchell/3581 (Greg Mitchell is author of the book *Atomic Cover-up*)

5. The Pauling Blog, Talking about the Limited Test Ban Treaty, http://paulingblog.wordpress.com/2013/10/30/talking-about-the-limited-test-ban-treaty/

6. Harvey Wasserman and Norman Solomon with Robert Alvarez and Eleanor Walters, *Killing Our Own – The Disaster of America's Experience with Atomic Radiation*, A Delta Book, 1982. Available for download at http://www.ratical.org/radiation/KillingOurOwn/

7. Gayle Greene, The Woman Who Knew Too Much: Alice Stewart and the Secrets of Radiation, University of Michigan Press, Reprint edition (July 31, 2001, Available for sale at http://www.amazon.com/The-Woman-Who-Knew-Much/dp/0472087835

8. Cindy Folkers, *US Radiation Panel Recognizes No Safe Radiation Dose*, Nuclear Information and Resource Service, World Information Service on Energy, July 2005. (A one-page article reprinted from The Nuclear Monitor) http://www.nirs.org/radiation/radtech/nosafedose072005.pdf

9. Book review article of Gayle Greene's book mentioned in note 7. http://www.jstor.org/discover/10.2307/3343289?uid=3739840&uid=2129&uid=2&uid=70&uid=4&uid=3739256&sid=21103332410961

Paul Zimmerman, *A Primer in the Art of Deception* is a good source of information on this topic. See http://www.du-deceptions.com/

10. GreenMedInfo.com - Why There Is No "Safe" Level Of Radiation from Fukushima, http://www.greenmedinfo.com/page/greenmedinfocom-why-there-no-safe-level-radiation-fukushima

11. Ernest Sternglass, Secret Fallout: Low-Level Radiation from Hiroshima to Three-Mile Island, McGraw-Hill Book Company, 1981, available for download at http://www.ratical.org/radiation/SecretFallout/SF.pdf

12. John W. Gofman and Arthur R. Tamplin, Poisoned Power: The Case Against Nuclear Power Plants Before and After Three Mile Island, Rodale Press, 1971 and 1979. Available for browsing at http://www.ratical.org/radiation/CNR/PP/

13. Interview with John W. Gofman titled *There is no safe threshold* in Synapse, Vol.38 No.16, January 20, 1994, University of California San Francisco. Available for download at http://www.ratical.org/radiation/CNR/synapse.pdf

14. Harvey Wasserman, People Died at Three Mile Island, https://www.commondreams.org/view/2009/03/24-3

15 See note 6. *Killing Our Own.*

16. Fukushima fallout much worse than Chernobyl, Update: Japanese children getting thyroid cancer, http://owndoc.com/radiation/fukushima-nuclear-fallout/

17. John W. Gofman, Radiation-Induced Cancer from Low-Dose Exposure – An Independent Analysis, 1990, Available for browsing at http://www.ratical.org/radiation/CNR/RIC/

18. Prof. Karl Grossman, Chernobyl Death Toll: 985,000, Mostly from Cancer, http://www.globalresearch.ca/new-book-concludes-chernobyl-death-toll-985-000-mostly-from-cancer/20908. The article refers to the book by Alexey V Yablokov, Vassily B. Nesterenko and Alexey V. Nesterenko, Chernobyl: Consequences of the Catastrophe for People and the Environment. The book is available for download at http://www.strahlentelex.de/Yablokov_Chernobyl_book.pdf

19. Nick Meo, Chernobyl's arch: Sealing off a radioactive sarcophagus, BBC News Magazine, 27 November 2013, http://www.bbc.com/news/magazine-25086097

20. Dr. David Suzuki, Filling in the Gaps on Fukushima Radiation and Its Effects on Fish, http://ecowatch.com/2014/01/28/fukushima-radiation-effects-fish/

21. Joseph J. Mangano and Janette D. Sherman, An Unexpected Mortality Increase in the United States Follows Arrival of the Radioactive Plume from Fukushima: Is there a Correlation? International Journal of Health Services, Volume 42, Number 1, Pages 47–64, 2012, © 2012, Baywood Publishing Co., Inc. Available for download at http://www.radiation.org/reading/pubs/HS42_1F.pdf

22. Harvey Wasserman, Toll Mounts Among U.S. Sailors Devastated by Fukushima Radiation, http://ecowatch.com/2014/01/11/sailors-devastated-by-fukushima-radiation/

23. Brandon Baker, 70+ USS Ronald Reagan Crew Members, half Suffering From Cancer, to Sue TEPCO of Fukushima Radiation Poisoning, http://ecowatch.com/2013/12/27/ronald-reagan-cancer-sue-tepco-fukushima-radiation/

24. See note 22.

25. Operation Tomodachi, Yokota Air Base, Japan, 7 April – 5 May 2011, Observations, Insights and Lessons. Available for download at http://info.publicintelligence.net/CALL-Tomodachi.pdf

26. Brandon Baker, Fukushima Nightmare Continues as Homeless People are Recruited for Cleanup, Scammed Out of Wages, http://ecowatch.com/2013/12/30/fukushima-homeless-people-recruited-scammed-wages/

27. Mike Adams, Fukushima now in a state of emergency, leaking 300 tons of radioactive water into the ocean daily, http://www.naturalnews.com/041610_fukushima_radioactive_leak_state_of_e mergency.html

28. Japan Reacts to Fukushima Crisis By Banning Journalism, http://www.washingtonsblog.com/2013/11/japan-reacts-fukushima-crisis-banning-journalism.html

29. Leading Scientist on Fukushima Radiation Hitting West Coast of North America: "No on is measuring so therefore we should be alarmed" – Federal, State and Local Governments *Refuse* to Test for Radiation on the West Coast of North America http://globalresearchreport.com/2014/01/27/leading-scientist-on-fukushima-radiation-hitting-west-coast-of-north-america-no-one-is-measuring-so-therefore-we-should-be-alarmed/?utm_source=feedburner&utm_medium=email&utm_campaign=Feed%3A+globalresearchreport+%28Global+Research+Report%29#sthash.Ntz9gLQ0.1a2BPRl0.dpbs

30. akasprak, The Revolution Will Be Clumpy, The Sieve, September 19, 2012, http://the-sieve.com/tag/clumped-isotopes/

31. Burkart W, Finch GL, Jung T, Quantifying health effects from the combined action of low-level radiation and other environmental agents: can new approaches solve the enigma? (Abstract) http://www.ncbi.nlm.nih.gov/pubmed/9352670

32. See note 20.

33. Fake Science Alert: Fukushima Radiation Can't Be Compared to Bananas or X-Rays, http://www.washingtonsblog.com/2013/04/fake-science-alert-fukushima-radiation-cant-be-compared-to-bananas-or-x-rays.html

34. Wikiquote, Hyman G. Rickover, http://en.wikiquote.org/wiki/Hyman_G._Rickover (see second quote).

What's the Business of Business?

Corporate responsibility and energy transition after the Fukushima crisis

Christian T. Lystbaek

Introduction

On March 11th 2011 a magnitude 9 earthquake hit northeast Japan. Less than one hour later a tsunami with waves of up to 40 meters struck the coast. The earthquake and the massive tsunami that followed triggered failures at four reactors at the Fukushima Daiichi nuclear power plant, causing explosions and serious radioactive leaks. Were the impacts on nuclear power stations preventable?

Earthquake risks are well known in Japan, and their occurrence is expected. Nowhere else in the world are seismic activity regions so congruent to nuclear power plant locations as in Japan. Thus, not only was a strong earthquake in Japan's northern region to be expected, it was overdue according to probabilistic calculations (Fujiwara, 2006). But if expected, would the electric power company operating the Fukushima Daiichi station, Tokyo Electric Power Company (TEPCO), the Japanese Nuclear Safety Commission (NCS) or the Nuclear and Industrial Safety Agency (NISA) have been able to prevent the accident? TEPCO had confirmed seismic safety of its nuclear power stations. Why did safety systems fail? Was it bad luck or was it complete ignorance of risks and safety rules? Whose responsibility was it?

There is plenty of reason to criticize TEPCO, its communication and management, both before and after the disaster in March 2011. By scanning the media one finds many articles criticizing TEPCO's communication and safety culture. TEPCO has a history of false reporting to the government and a lack of transparency towards the public. In 2002, a whistleblower revealed systematic concealment of plant safety incidents and false reporting in routine governmental inspection of its nuclear plants. And again on March 2nd 2011, only nine days before the Fukushima Daiichi meltdown, NISA found that TEPCO had violated safety regulations and instructed the company to put in place preventive measures. After the meltdown on March 11th, the crisis information from the company was slow, chaotic and mistrusted by the media. Further, it became clear that TEPCO lacked adequate disaster prevention and mitigation plans. TEPCO had to ask for volunteers, of which they only found a few, and therefore had to hire temporary employees. Thus, there is plenty of reason to criticize TEPCO. Today, TEPCO accepts this criticism and apologizes on its webpage for not living up to its responsibilities.

In this chapter, I will address the question of corporate social responsibility (CSR) in regards to the Fukushima crisis. I will not, however, discuss "who is to blame?" for the Fukushima Daiichi power plant meltdown. Rather, I will discuss corporate responsibility in regards to the energy transition toward a sustainable energy system.

Today, the energy system is changing toward sustainable development by means of renewable energy and energy efficiency. This energy transition is driven primarily by concerns about climate change, resource availability and sustainability. As many scholars argue, important environmental issues cannot adequately be dealt with without considering the crucial roles of corporations: Corporations are crucial players in terms of economic force, political power and environmental influence (e.g. Bendell & Bendell, 2007). In response to this, corporations have increasingly articulated social and environmental responsibility. This has given rise to a new form of corporate eco-communication, distinct from the crisis communication arising out of environmental disasters in earlier decades (Cheney et al., 2007). This new kind of eco-talk has involved businesses in adopting and adapting emerging concepts of CSR and sustainable development in order for them to express concern about the environmental impact of their activities, for instance in regards to energy efficiency and energy transition in general. Some energy-intensive companies have constructed electricity and heat-generation plants to meet their own energy needs.

Critical scholars and environmentalists, however, claim that much of the eco-talk is just "greenwashing", that is, environmentally friendly rhetoric without real substance. Green advertising and CSR in general has become a management fashion which is used as a potential source of marketing advantage for companies. Consequently, many companies are allegedly supporting environmental activities, despite consumer skepticism about their environmental claims (Zorn & Collins, 2007). Several scholars and NGOs have argued that the nuclear industry is among the worst "greenwashers", since it attempts to rebrand nuclear power as "clean" despite its contaminating character (Ongkrutraksa, 2007).

In the following, I will address the question of CSR in regards to energy transition. Firstly, I will analyze the CSR framework at TEPCO as this is described in corporate documents and on the company webpage. Secondly, I will analyze the corporate social responsibility policy framework in Japan in general.

CSR and Energy Transition at TEPCO

Japan is divided into regions, each of which receives electricity service from a regional provider. In other words, the regional electric power companies hold a monopoly of generating and distributing power within their designated region. The electric power company operating the Fukushima Daiichi station was TEPCO, established in 1951 to supply electricity to the Kanto region, including metropolitan Tokyo. Thus, TEPCO's service area is home to approximately one-third of Japan' population and TEPCO's electricity sales represent approximately one-third of total electricity sales in Japan, putting the company on a level with the world's major electric power companies. Before the disaster in March 2011, TEPCO operated three nuclear power stations – the six-reactor Fukushima Daiichi station, the four-reactor Fukushima Daini station and the seven-reactor Kashiwazaki-Kariwa station in Niigata Prefecture.

TEPCO has several documents describing its CSR activities and achievements, including a Sustainability Report (TEPCO, 2010) and webpage information about its "Corporate Ethics", "Efforts to promote renewable energy" and how it develops its "power supply facilities" in general. In these documents TEPCO articulates commitment to CSR and sustainable energy transition at strategic, policy and programmatic levels, that is, in descriptions of TEPCO's strategic position, in corporate policies and in CSR programs such as mission statement, planning documents, management systems and programmatic guidelines. For instance, the Sustainability Report starts with a mission statement, which states that:

> The TEPCO Group's basic mission is to deliver electricity to society in a safe and stable manner. Based on this commitment, we will pursue new social and environmental roles in 'leading the low-carbon era' as we contribute to creating affluent and comfortable life-styles. [...] Our idea about affluent and comfortable environments is a sustainable society that is not only convenient and comfortable to live in, but that is filled with spiritual affluence and harmonizes with nature. (TEPCO, 2010:4-6)

Thus, TEPCO articulates commitment to CSR and sustainable development at a strategic level. Further, at a programmatic level, the Sustainability Report states that

> The TEPCO Group will [...] pursue new social and environmental initiatives. We will make active and integrated efforts to reduce carbon from all aspects of the energy chain, from the power generation to utilization stages, or from the electricity supply side to demand side. (TEPCO, 2010:5).

Again, such programmatic guidelines articulate a commitment to CSR and energy transition. This programmatic commitment is also found on TEPCO's English webpage (http://www.tepco.co.jp/en/index-e.html > Power Supply Facilities > Nuclear). Under the heading "Energy and Resources. Nuclear", the webpage states that TEPCO "strives continually to develop and construct the electric power facilities needed to deliver the required electricity in a stable, economic, and environmentally friendly way". Contrary to the potential shift in momentum away from nuclear power created by the Fukushima power plant meltdown, however, TEPCO continues to promote nuclear power. It describes nuclear power as a "clean" and "emission-free" solution despite the ongoing environmental problems associated with it. Although TEPCO has introduced some power generation facilities that leverage renewable energy, such as solar, wind and hydropower, this is only to a limited degree.

The commitment to nuclear power is evident in TEPCO´s CSR documents. In the Sustainability Report, TEPCO notes that "[w]e conduct our business by operating nuclear plants, transmission and distribution lines, and a host of other facilities" (TEPCO, 2010:5). This commitment to nuclear power is described in more detail later in the Sustainability Report when TEPCO notes that "the best mix of power sources" is a balance of nuclear power, thermal power and hydropower (TEPCO, 2010:18). In this mix, nuclear power is presented as "a key power source" and the base supply, since "[n]uclear power generation is an outstanding generation method in terms of stability, environmental performance, and economic efficiency" (TEPCO, 2010:18). Thus, TEPCO regards nuclear as a solution, or as part of the solution, for a sustainable energy supply. Nuclear power is being presented as the best way to generate electricity without greenhouse gas emissions.

This commitment to nuclear power is also articulated on the TEPCO webpage. Under the heading "Power Supply Facilities", TEPCO states that it regards nuclear power as a sustainable, eco-friendly power source: "Nuclear power generation has excellent long-term prospects for the stable procurement of nuclear fuel and for effectively countering global warming problems." Under the subheading "Energy and Resources. Nuclear", the webpage states that "[t]o ensure a stable energy supply in the future, and to preserve the environment, TEPCO […] is promoting the use of nuclear power generation." Thus, the way in which TEPCO is planning to fulfill its CSR and sustainability commitment is through a commitment to nuclear power. Nuclear power is being put forward as an environmental friendly solution despite the ongoing environmental problems associated with it.

This commitment to nuclear power is also backed up by the Japanese government. On 14 September 2012, the Japanese government decided to phase out nuclear power by the 2030s, or 2040 at the very latest. The government said that it would take "all possible measures" to achieve this goal. A few days later the government retrenched the planned nuclear phaseout after the industry pushed for reconsideration. The government now argued that a nuclear phaseout would burden the economy, and that imports of oil, coal, and gas would bring high added costs. The government then approved the energy transition, but left open the time-frame for decommissioning the nuclear power plants.

Thus, contrary to the potential shift in momentum away from nuclear power created by the Fukushima crisis, nuclear energy continues to attract significant attention among energy companies, politicians and the media. Even some academics have argued for aggressive expansion of nuclear power to stave off climate change. Unfortunately, however, these views are naïve. Although TEPCO and others promote nuclear energy as sustainable, a nuclear accident is certainly not a sustainable event. The damage to the environment is huge.

Radioactive iodine from the Fukushima plant has been detected in the water supply of Tokyo, more than 220 kilometers to the south of the power plant (Pearce, 2012). Apart from radioactive emissions, it is simply wrong to consider nuclear power as a zero-emission source, also with regard to CO_2. Nuclear power is not "clean". It produces greenhouse gases and toxic and radioactive by-products. But despite these problems, nuclear power continues to attract significant attention.

Sovacool & Brossmann (2010; 2013) have argued that an explanation for the significant attention seems to be the way that nuclear power fulfills socio-psychological and cultural needs related to a future world where energy is abundant and pollution-free, an image or vision that manifests itself with the belief that society can continue to operate without limits imposed by population growth and destruction of the environment. This vision plays a powerful role in how people embrace a particular form of energy, such as nuclear power: "In short, how people imagine energy technologies and their futures is clearly important to understanding how and why people invest in them, financially, personally, professionally, and otherwise, and it is thus a critical social facet of energy transitions" (Sovacool & Brossmann, 2013:211).

Rhetorical Visions about Nuclear Energy

According to Sovacool & Brossmann, the compelling images of nuclear power as abundant and pollution-free serve as "rhetorical visions" (Sovacool & Brossmann, 2010) by means of which nuclear power continues to attract significant attention. Based upon a comprehensive literature review and a large number of interviews with energy experts, Sovacool & Brossmann have identified the rhetorical visions typically used to support a particular form of energy source, such a nuclear power. The most dominant and strong rhetorical visions are "inevitability" and "necessity". These visions are frequently represented in the documents describing TEPCO's CSR activities and achievements.

The rhetorical vision of inevitability expresses the belief that nuclear power is inevitable, that is, the only sustainable energy source (Sovacool & Brossmann, 2010:2005). This vision is often stated in both a historical perspective and an environmental perspective. In a historical perspective, it is noted that nuclear power is inevitable because of future shortages in alternative sources of power. This perspective is articulated repeatedly in TEPCO's documents. For instance, the Sustainability Reports notes that "[i]f the world continues to produce energy at the current rate, [...] resources will dry out" (TEPCO, 2010:16). Continuing along these lines, the webpage states that:

> [th]e world's population is predicted to continue increasing in the 21st century. Because of this population growth, combined with economic development in Asia and other regions, it is expected that global energy consumption will keep growing and

that supplies of energy reserves, which are limited, will become tight. [...]In such a situation, nuclear power is an indispensable energy source for ensuring a stable supply of energy and for responding to global warming issues." According to TEPCO, then, nuclear power is inevitable, because shortages in the conventional fossils fuels will make it so.

Further, TEPCO states that nuclear power is inevitable in an environmental perspective because it is environmentally sustainable. For instance, the Sustainability Report notes that "[n]uclear power generation is [...] environment-friendly. Because it uses the heat that is given off by uranium during nuclear fission, it releases no CO2, the primary cause of global warming, nor NOx and SOx, the major sources of air pollution, in the process of generating power." (TEPCO, 2010:18). And the TEPCO webpage notes that "[i]ncreasing energy consumption has brought up a variety of environmental issues. In particular, CO2 emissions from fossil fuels are thought to be one of the causes of global warming. Nuclear power does not emit CO2 in the generation process, making it the preferred option as an energy source for mitigating global warming." According to TEPCO, then, nuclear power is inevitable in an environmental perspective, because it excels in global environmental protection, due to the absence of atmospheric pollutants that cause global warming and acid rain.

As these statements make clear, the argumentative function of the rhetorical vision of inevitability has important implications for discussions about energy transition. For if nuclear power is inevitable, then the only obstacles are transitory and opposition is ultimately fruitless. Thus, this vision calls upon politicians and the public to accept nuclear power in an unprejudiced manner, and completely erases the technical and environmental challenges.

The rhetorical vision of "necessity" expresses the belief that nuclear power is necessary. Again this belief is presented in different perspectives; both regarding economic growth and regarding energy independence (Sovacool & Brossmann, 2010:2006). The point about economic growth centers on the claim that nuclear power is economically effective and necessary in order to maintain economic development. This point of view is also frequently articulated in TEPCO's documents. For instance, the Sustainability Report notes that nuclear power is characterized by "outstanding economic efficiency unaffected by fuel price changes." (TEPCO, 2010:18). The TEPCO webpage also notes that "[a]s a result of worldwide population and economic growth, energy consumption is expected to increase greatly in the 21st century." In this situation, nuclear power is the most cost-effective of energy sources: "[n]uclear power generation produces massive amounts of energy from comparatively small quantities of fuel. Once fuel has been added to the reactor, a nuclear power plant can run continuously for approximately one year without adding or exchanging fuel." According to TEPCO, then, nuclear power is necessary in order to maintain current levels of consumption and economic growth in the industrialized and industrializing world.

Further, it is stated that nuclear power is necessary in order to enhance energy independence and energy security. In the Sustainability Report, TEPCO notes that:

> [a]lthough Japan is the third largest consumer of energy in the world, its sparse energy resources create dependence on overseas supplies [...]. TEPCO has striven to become more energy-secure by, for instance, diversifying both the types of energy and their points of import as far as possible. [...]To ensure a stable energy supply in the future, and to preserve the environment, TEPCO, committed to its safety-first principle, is promoting the use of nuclear power generation."

Even though TEPCO is considering nuclear power "a semi-domestically produced energy" (TEPCO, 2010:16), Japan is dependent on imports of uranium, primarily from Australia and Canada. Thus, nuclear power is seen as a way to minimize the dependence on foreign sources of fuel, all the while contributing to energy security.

As these statements make clear, the argumentative function of the rhetorical vision of necessity adds to implications for discussions about energy transition. For if nuclear power is necessary for the growth and independence of the economy and, ultimately, the society, then critical comments can be presented as critical towards the economy or society in general. Hence, this rhetorical vision is easily associated with patriotism whereby nuclear power is presented as the only a way to renew the independence and security of the society (Sovacool & Brossmann, 2010:2006). Eisako Sato, governor of the prefecture of Fukushima from 1988 to 2006, has noted how this has happened in Japan. Because Japan is believed to be dependent on nuclear energy, it has been legitimate for officials to hide incidents in order to avoid bringing trouble to the country, and opponents of nuclear energy generation have been treated like public enemies. Sato describes to this as "nuclear absolutism" (Koh, 2011).

Taken together, the argumentative function of the rhetorical visions of inevitability and necessity cannot be overstated; the inevitability aspect enables proponents to reject counterarguments and the necessity aspect enables them to justify calls for a major commitment from public and private actors. The implication for discussions about nuclear power is that the challenges faced by nuclear energy are frequently discounted in the face of much more compelling (and exciting) visions. Consequently, discussions about nuclear power often transcend rationality and logic since they involve elements as diffuse as communal hope, economic optimism, social security and national pride (Sovacool & Brossmann, 2013:211).

But nuclear power is neither inevitable nor necessary. Rather, these rhetorical visions express an attempt to legitimize or "greenwash" nuclear power by rebranding it as inevitable and necessary despite its contaminating character.

The Greenwashing of Nuclear Energy

"Greenwash" (a compound word modeled on "whitewash") entered the official lexicon of the English language in the 1990s through its inclusion in the *Oxford English Dictionary*, which defines the term as: "Disinformation disseminated by an organization so as to present an environmentally responsible public image." Hence, "greenwashing" is a form of spin in which green PR or green marketing is deceptively used to promote the perception that an organization's aims and policies are environmentally friendly. The term is generally used when significantly more money or time has been spent advertising being *green*, that is, operating with consideration for the environment, rather than spending resources on environmentally sound practices (Ongkrutraksa, 2007). Energy companies, which are traditionally the largest polluters, have long been criticized for "greenwashing". Especially the nuclear industry has been criticized for attempts to "greenwash" nuclear power by rebranding it "clean" and "emission free" despite its contaminating character. The Advertising Standards Canada, the organisation which regulates the Canadian advertising industry, ruled that claims of nuclear power being "emission free" made in adverts "were inaccurate, unsupported, and misleading" (Ongkrutraksa, 2007). Fossil fuels are essential to every stage of the nuclear cycle, and CO_2 is given off whenever these are used. The mining and processing of uranium and the construction of the power plants generate significant greenhouse gas emissions. Nuclear power is simply not greenhouse gas emissions free. Rather, when viewed over the nuclear fuel life cycle, nuclear greenhouse gas emissions can be substantial (Pearce, 2012).

Nuclear power, then, is not the inevitable environmentally friendly solution to energy transition. Although common in nuclear interest groups and even in peer reviewed literature, it is a misperception that because nuclear energy does not produce carbon dioxide as a byproduct during electricity generation, as do fossil fuels, it is an "environmental friendly" source of energy. It produces toxic and radioactive by-products. Every kilowatt-hour of nuclear energy is responsible for CO_2 emissions; the amount depends on the location and a number of technical factors. For example, depending on the source and quality of the ore, milling, mining and transporting of uranium, some studies have shown nuclear plants can even emit about the same amount of CO_2 per unit of electricity as a natural gas power plant (Pearce, 2012). Thus, nuclear power is not the inevitable environmentally friendly solution to the energy transition toward a sustainable energy system.

Nor is nuclear power inevitable because of a shortage of alternatives. Although abandoning nuclear energy will not realize the hopes of many stakeholders by completely replacing this form of generation with renewables in a very short run, renewable energies can be expanded and integrated into the energy system in order to achieve the targets for reduction of CO_2 emissions. Up to a few years ago, electricity in the industrialized part of the world was primarily generated in large-scale coal-fired, gas-fired and nuclear power stations. Over the course of time, new power stations with ever greater capacity were constructed with the aim of making the most of the efficiency benefits provided by bigger units. But the economic foundation underpinning the generation of electricity with fossil fuels has meanwhile been increasingly dismantled and renewable energy sources, such as wind, biomass, hydropower and solar power, now serve as alternatives. Investments in offshore or onshore wind farms also give rise to larger generating units. However, the proportion of smaller units for electricity production is rising at the same time, for example with the installation of solar panels on the roofs of residential accommodation. Many energy-intensive companies have constructed electricity and heat-generation plants to meet their own energy needs. Apart from efficiency and cost awareness, the drive towards decentralization, supplier diversity and self-determination have become important benchmarks for issues relating to the supply of energy. Many players are demanding greater involvement in the generation of electricity and they are accepting an increase in energy costs as a consequence. This change in the energy landscape is structural rather than temporary, and the main instrument for making the energy supply more climate friendly is the expansion of renewable energies. Thus, nuclear power is not the inevitable solution to shortages of fossils fuels.

Further, nuclear energy is neither necessary in order to maintain economic growth nor to minimize dependence on foreign sources of fuel. First of all, it is a misrepresentation to promote nuclear energy as a means to minimize dependence. Japan has very few conventional energy resources of its own. Its self-sufficiency for energy is around five percent, which is substantially lower than other major industrialized countries. But even though TEPCO is willing to consider nuclear power "a semi-domestically produced energy" (TEPCO, 2010:16), Japan is dependent on imports of uranium, primarily from Australia and Canada. Thus, nuclear power does not reduce the dependency on foreign sources of fuel.

It is also a misrepresentation to promote nuclear energy as a means to maintain economic growth. The relatively low nuclear electricity prices stem from the fact that not all costs and risks are priced in, that nuclear power production is strongly subsidized and external costs are not internalized. Economists discuss this in terms of "externality" as a cost not transmitted through prices. Nuclear power has a long list of externalities, including environmental externalities and externalities imposed on future generations. It is beyond the scope of this chapter to go into each in detail, however, some of the fundamental externality challenges to using nuclear energy for widespread electric generation are, first, that nuclear power is dependent on the mining of a finite uranium rich ore. Much like the mining and processing of other materials, uranium mining, processing and enrichment can leave substantial damage to the nearby ecosystems and waterways (Pearce, 2012). At the end of the life cycle there are externalities that continue for generations. Safe and secure ultra-long-term storage of nuclear waste is still unresolved and poses a serious challenge given the half-life of spent fuel. This waste containment problem includes not only the spent fuel rods, but also upon decommissioning of a nuclear power plant, the building, equipment and the surrounding land also contribute to the total waste.

Both current temporary storage methods and planned methods of burying underground are inherently inequitable as those who bear the consequences – including future generations and socio-economically disadvantaged people living near facilities – have little or no say in the decisions (Pearce, 2012).

A larger and more serious externality during operation is nuclear insecurity. No matter how small the probability, safety concerns with nuclear power plants pose a real and finite danger. The potential for a nuclear disaster imposes risk related costs to essentially everyone in the world. An aging worldwide reactor fleet might be expected to become less reliable, and a recent review of the accident record questions the accepted assumption that new designs are less risky (Sovacool, 2011). These risks of very large-scale problems cannot be eliminated entirely, as was made very clear by the Fukushima disaster. Risk factors with nuclear power include potential nuclear disasters due to accidents, negligence, poor design, natural disasters, terrorism and the threat multiplication possible with nuclear energy in operation during both international wartime and domestic conflicts. Thus, continued use and development of nuclear technology carries serious proliferation risks (Pearce, 2012). All these externalities – and more, such as human health effects, biodiversity loss, land degradation, diverse social costs, etc. – are very difficult to quantify. Thus, nuclear energy is not necessary in order to maintain economic growth. One could argue that there is a market failure when the market fails to account for these externalities. But it seems more reasonable to argue that it is not the market that fails to account for the costs of nuclear power, but that the cost of nuclear power fails to qualify for a market, because the costs are too big and uncontrollable to be managed. One could argue, then, that nuclear power is not suitable for a society that values sustainability in economic, social and environmental terms.

In the remainder of this chapter, I will analyze the CSR policy framework in Japan. As the analysis will make clear, special efforts are required in order to expand renewable energies and integrate these capacities into the energy system.

CSR in Japan – and in General

During the 1990, much of the world was attracted to Japanese innovative management practices, such as total quality management and just-in-time management. These advanced management practices enabled the production and distribution of high-quality goods and services. Furthermore, workplace safety and health promotion have become a benchmark for many companies. In addition, life-time employment and extensive benefit programs in Japanese companies resulted in strong loyalty and high morale of employees (Wokutch & Shepard, 1999). Thus, the CSR performance of Japanese companies with regards to the work environment is remarkable compared to other Asian countries, but still, the CSR performance of Japanese companies in regards to energy transition has not met the expectations of many scholars (Chapple & Moon, 2005). Why has Japan failed to meet the expectations in the field of CSR and energy transition?

Scholars seem to agree that a main reason is the historical division of power between government and companies in Japan (Lewin et al., 1995; Wokutch & Shepard, 1999). Government and companies play their own distinct roles without interfering with each other. This has restricted the influence of government on CSR. Rather than direct government intervention in Japanese business activities by the means of regulation or indirect government intervention by means of market-based instruments, the Japanese government has relied heavily on voluntary agreements and guidelines. For instance, the government published a report in 2005 on corporate social responsibility entitled *Sustainable Environment and Economy*, which proposes ideal images of sustainable business, and another in 2012 entitled *Environmental Report Guideline to Realize a Sustainable Society*. Furthermore, when it establishes guidelines for CSR, the government collaborates closely with the Japanese Business Federation, Keidanren, and takes into account the opinion of industry. For instance, in 2005, METI adopted Keidanren's proposal to corporate social responsibilities with very little alteration. As a result, most of the government guidance has not exceeded the expectations and control of the companies themselves.

This over-reliance of the Japanese government on administrative guidance fails to establish institutionalized CSR programs and systems. Government guidelines result in confusion about the range and definition of CSR. Although legislation restricts the freedom and flexibility of business, it establishes clear standards and agreements on controversial issues. The Japanese government has not created enough internal societal pressure to motivate companies to develop CSR practices that exceed limited and restricted CSR activities. Nor have Japanese customers. Japanese consumers have experienced a steady rise in economic prosperity and fulfillment of basic economic needs. In recent years, this has brought about greater demands for CSR by Japanese citizens, and Japanese consumers prefer to support companies which have a good CSR track record. Survey data show that 62.4 percent of respondents will give priority to buying from companies that are socially responsible and have a sound ethical policy (METI, 2004). Nevertheless, it seems doubtful that the strong interest of Japanese consumers in CSR generates enough societal pressure to affect the CSR agenda of Japanese companies. Traditionally, the consumer movement in Japan has been relatively weak. Furthermore, Japanese companies place little importance on consumer interests due to their protectionist trade policies, which place industry interests over consumer interests (Lewin et al., 1995; Wokutch & Shepard, 1999). Therefore, it is uncertain how many consumers actually carry over their intention to support CSR with their behavior, such as boycotting the product and services of irresponsible companies.

Consequently, the Japanese government, as well as the nuclear industry and Japanese companies in general, must make active and integrated efforts to promote the transition toward a sustainable energy system.

A Narrow, Economic Conception of CSR

Without close intervention of the government or other CSR actors, Japanese companies have tended focus their CSR activities on CSR areas that are directly aligned with their profitability, most notably employee health. For long, Japanese companies have enjoyed autonomy in formulating CSR strategies without intervention or control from domestic and international actors. As a result, many Japanese companies have only given priority to CSR areas in which they can see a clear connection to profit and return on investment (Lewin et al., 1995). For example, many companies have recognized that improving the working environment and the active participation of workers increase employees' commitment to their work, which results in boosts to worker productivity and loyalty, as well as to product quality. Furthermore, the focus on occupational health and safety has reduced productivity loss caused by illness and accidents. But traditionally, many Japanese companies have perceived CSR as an instrument to increase their profits and to generate competitive advantage. As a result, they have made progress in certain areas which align with their self-interest, such as working conditions (Wokutch & Shepard, 1999). Thus, traditionally, such a narrow economic conception of CSR has been widespread among Japanese companies. Dunphy and colleagues (2014) have developed an analytical framework of different stances towards CSR. Applying this framework to the narrow economic conception of CSR, the responses to CSR in many Japanese companies range from non-responsiveness to active rejection.

Non-responsiveness is characterized by a lack of awareness or interests other than immediate financial viability. When taking this stance, the environmental consequences of corporate behavior are simply outside the scope and not on the agenda. Traditionally, this has been the stance in many companies, which typically have not addressed environmental issues (Crane & Matten, 2010), also in Japan, where most companies have considered CSR as a passive response to pressures from the market (Tanimoto, 2010). This is particularly significant in smaller companies that are not required by law to comply with environmental regulations. The logic that supports non-responsiveness is that restrictions should not apply to those smaller companies that cannot afford to comply. In this case, the ability to make a profit must be treated as most important. The same logic precludes smaller companies from having to comply with laws regarding socially responsible legislation.

Another, more active response is rejection of corporate responsibilities beyond financial ones. Rather than neglecting a concern for the natural environment, this stance entails an active rejection of responsibilities beyond economic gain within the formal bounds of society. This stance is warranted from the neo-liberal economic view of the often-cited Milton Friedman (1962). According to Friedman, corporations, or more realistically corporate executives, are obliged to owners and must act in the owner's interest. To the extent that they do not marshal available resources toward this end, they are not functioning optimally or responsibly. At first glance, this view may seem somewhat extreme and anachronistic, however, Japanese corporate law states that corporate management has duties of loyalty, which, if violated, can be grounds for shareholder lawsuits. In this case, the prevailing view among many economists and corporate directors is, still, that making a profit must be treated as most important for companies.

Whether in terms of non-responsiveness or rejection, the narrow economic conception of CSR is still widespread, in Japan and elsewhere. Without close intervention or control from the government, Japanese companies have tended to prioritize CSR areas with a clear connection to profit. The routine corporate focus on "efficiency" and "the bottom line" are examples of how corporations may be encouraged to adopt non-responsiveness or rejection. Companies exist within a context that requires profitability. Thus, their most significant focus is the pursuit of profit. The one-sided focus on profit in this narrow economic conception of CRS is grounded in economical and organizational traditions of taking the environment for granted (Livesey & Graham, 2007). Employing a utilitarian view, the natural environment is seen as available resources, which are not accounted for but are, rather, treated as externalities. Although many Japanese companies are publishing sustainability reports, such reports, like TEPCO's report, are mostly concerned with how CSR is structured within the organization, and less concerned with specific activities (Kolk, 2008; Tanimoto, 2013). Thus, there is a huge risk of "greenwashing" in these reports.

A Broader Economic Discourse on Sustainability and its Implications

However, this narrow economic conception of sustainability is being increasingly criticized. Many policy initiatives at local, regional, national and international levels have been encouraging broader conceptions of CSR. In particular, the United Nations World Commission on Economic Development (known as the Brundtland Commission), defining stainable development as the ability to "meet the needs of the present generation without compromising the ability of future generations to meet their own needs" (UNWCED, 1987:8), contributed to the introduction of broader conceptions of sustainable development into politics and policies. Using scientific evidence to mobilize public awareness of the toxic side effects of industrial progress, the Brundtland Commission showed the risk of unproblematic commitment to economic development and called for an approach balancing different needs and risks, costs and benefits. Since 2000, the United Nations has urged governments and corporations to embrace the *Global Compact*, a call to governments and companies around the world to take action in regards to the development, implementation, and disclosure of responsible corporate policies and practices: "The *Sustainable Energy for All* Initiative will call for all partners to take bold action – through strengthened enabling policy, transformative partnerships and collaboration, enhanced financing, and new innovation. All partners have a role to play. The Initiative will seek significant new commitments and policy changes at the government and national level. Additionally, it will encourage the private sector to drive investment, increase innovation in products and services, and increase operational efficiencies. To do this, it is imperative that executive leadership in the private sector commit to the Initiative and make access to energy, energy efficiency, and renewable energy a strategic priority across their organizations, from the boardroom to the facility level." (UN, 2011:2) Thus, accepting that companies must make a profit, this view states that this judicial record leaves room for the possibility that companies may sacrifice some profit in the public interest. Applying the analytical framework of different stances toward CSR developed by Dunphy and colleagues (2014), this broader conception of CSR encourages a stance toward CSR which ranges from compliance to proactivity.

Compliance involves awareness that bad publicity, community action and legal sanctions can be harmful to the corporation's bottom line. Compliance may involve both adapting to external pressures and attempting to control these pressures. Companies that are compliant, then, follow legal constraints and attempt to adapt to stakeholder expectations to avoid undue criticism. At a minimum, compliance is encouraged through the desire to avoid the financial cost of non-compliance, but perhaps more importantly, the reputational costs of non-compliance. Although compliance can be very costly for corporations, since compliance often entails hiring specialists and maybe even the need for more substantial structures such as departments and information systems, these costs may by far outweigh the cost of non-compliance.

Another stance, which does not assume the environment to be a problem or cost, but more positively sees concern for the environment as an opportunity or a saving, is proactivity. Proactivity is the stance that the benefits of extended environmental responsibilities outweigh cost, or even can be turned into a win-win, i.e. a strategic business opportunity and competitive advantage. This stance represents a shift from a more defensive posture to an assumption that an environmental focus can be a positive benefit for the organization. Rather than viewing an environmental focus as a cost or imposition, a company can realize positive gains from its environmental attention. This is, then, a win-win proactive stance. A company may develop innovative products or services for new markets or strategize to position itself as an environmental leader in an established market. Clearly, this stance is feasible for organizations to the extent that they are able to identify and benefit from opportunities to improve efficiencies. They are able to continually improve their use of energy, waste and operations so that the natural environment and their profits both benefit (Dunphy et al., 2014:95f).

With increasing recognition of environmental problems, some companies are beginning to adopt such a broader conception of CSR and sustainability, either in terms of compliance or proactivity. They are recognizing that advanced environmental CSR activities give them a competitive advantage over other companies. Some companies have realized huge savings as waste has been reduced and efficient systems introduced. They have saved both resources and money, for instance by adding heating insulation to buildings and purchasing more fuel efficient-vehicles. Thus, efforts to reduce environmental burdens have led to greater efficiency, lower costs and the creation of new markets for environmentally friendly goods and services through innovation in technology and management (Fukukawa & Moon, 2004). The Japanese government has to some extent encouraged such efforts. The Japanese government has committed Japan to emission reduction under the Kyoto Protocol, and in recent years, it has started to use market-based instruments to promote this commitment. For instance, since 2008, the governmental approach has started to reduce taxes for ecological products which contribute to the reduction of greenhouse gas emissions as a part of economic stimulus policy. Thus, low emission vehicles such as hybrid cars and low-fuel consumption cars have been the subjects of lowering tax rates.

A "Deep", Systemic Discourse of Sustainability and its Implications

According to some scholars, even this "broad" conception of CSR is considered insufficient to deal with environmental problems. As long as growth is primary, one can imagine an end to needed natural resources or a point where the lack of clean air, water, and food brings about a collapse of human societies. In Japan, Kanji Tanimoto (2004, 2010, 2013) has argued that voluntary and incentive-based efforts of individual companies are insufficient to promote sustainable development of the whole socio-economic system. Rational activity by an individual does not necessarily heighten the rationality of society as a whole (Tanimoto, 2010). So even if each company were to tackle CSR issues individually, such efforts do not guarantee a contribution to the sustainable development of society as a whole. A system that relies only upon the spontaneity of companies will only result in those companies doing 'what they can' or 'what they want' on various CSR issues. As such, companies tend to focus on philanthropic activities that have public appeal, rather than to make efforts to enforce actual CSR progress. Despite the boom in CSR reporting, if each company individually sets their own standards of trustworthiness and transparency, they cannot be tracked easily.

One can find elements of this conception in the Brundtland report and in current United Nations' initiatives in regards to Sustainable Energy for All (e.g. UN, 2012). These UN documents argue that sustainability implies an enhanced understanding of environmental concerns as systemic issues that needs to be addressed in new ways. The best interests of single corporations are not the sole focus. Instead, an understanding of the implications of the environmental crisis leads to an understanding that the wellbeing of all people and the biological systems that support them are mediated through corporate activities. This systemic conception of sustainability highlights the embedded nature of the organization within the economic, social, legal and environmental spheres, each of which generates corporate responsibilities. Thus, it serves to draw attention to the networks in which the organization partakes and without which it could not exist.

In Japan, Tanimoto insists that different kinds of "rules" are necessary to support sustainable development. Voluntary rules by industrial associations and ethical norms by NGOs have performed moderate roles in setting a framework, but public policy plays a significant and indispensable role in creating an overall framework for companies to readily address CSR issues, and to encourage the development of CSR in the marketplace (Tanimoto, 2010). Public policy plays several important roles in the enhancement of CSR. Firstly, public policy creates a legal framework to control business, for instance in order to protect consumers through the regulation of toxic substances. Secondly, public policy creates institutionalized fiscal frameworks, which serve to encourage corporations to enhance CSR, for instance by setting standardized guidelines for CSR-reports, promoting CSR benchmarking, granting awards to socially responsible companies, etc. Thirdly, public policy creates collaborations between government, business and NGOs on CSR issues. It is not possible for government alone to mainstream CSR in market society. It is important to collaborate with other sectors to formulate CSR policies, and to address local community development (Tanimoto, 2010). Consequently, a "deep", systemic perspective is needed to embed CSR in Japan, which involves policy tasks at both macro, intermediate and micro levels. At the macro level, the Japanese government is required to make institutionalized legal and fiscal frameworks to promote CSR. At the intermediate level, policy can promote collaboration between industrial associations, labor unions, consumer groups, NGOs and research institutes. These actors play crucial roles in the development of CSR in the market society. At the micro-level, the individual corporation needs to incorporate CSR into their business and management plans and practices. Companies can obtain improved trust and enhance corporate reputation through the positive disclosure of non-financial data about environmental issues (Tanimoto, 2010).

Concluding Remarks

In this article, I have addressed the question of CSR in regards to energy transition after the Fukushima crisis. I have analyzed the CSR framework at TEPCO as this is described in corporate documents and the CSR framework in Japan in general in the light of different stances toward CSR.

The analyses show that, contrary to the potential shift in momentum away from nuclear power created by the Fukushima power plant meltdown, TEPCO continues to promote nuclear power. It describes nuclear power as a "clean" and "emission-free" solution despite the ongoing environmental problems associated with it. But this is a misrepresentation. A nuclear accident is certainly not a sustainable event; the damage to the environment is huge. Apart from radioactive emissions, it is simply wrong to consider nuclear power as zero-emission source, also with regard to CO_2. Nuclear power is not "clean". It produces greenhouse gases and toxic and radioactive by-products. But despite these problems, nuclear power continues to attract significant attention.

An explanation for this significant attention seems to be the way that nuclear power fulfills socio-psychological and cultural needs related to a future world where energy is abundant and pollution-free and society can continue to operate without limits imposed by population growth and destruction of the environment. This need manifests itself in the dominant and strong rhetorical visions about "inevitability" and "necessity". These visions are frequently represented in the documents describing TEPCO's CSR activities and achievements. Taken together, the argumentative function of the rhetorical visions of inevitability and necessity cannot be overstated; the inevitability aspect enables proponents to reject counterarguments and the necessity aspect enables them to justify calls for a major commitment from public and private actors. The implication for discussions about nuclear power is that the challenges faced by nuclear energy are frequently discounted in the face of much more compelling (and exciting) visions. Consequently, discussions about nuclear power often transcend rationality and logic since they involve elements as diffuse as communal hope, economic optimism, social security and national pride. But nuclear power is neither inevitable nor necessary. Rather, these rhetorical visions express an attempt to legitimize or "greenwash" nuclear power by rebranding it as inevitable and necessary despite its contaminating character. Although nuclear interest groups and even peer reviewed literature describe nuclear power as "clean" and "emission-free", this is a misrepresentation. Nuclear energy produces toxic and radioactive by-products. And every kilowatt-hour of nuclear energy is responsible for CO_2 emissions. Thus, continued use and development of nuclear technology carry serious technological, social, environmental and economic costs, many of which are very difficult to quantify. One could argue that there is a market failure when the market fails to account for these externalities, but it seems more reasonable to argue that it is not the market that fails to account for the costs of nuclear power, but the cost of nuclear power fails to qualify for a market. The costs are too big and uncontrollable to be measured and managed. One could argue, then, that nuclear power is not suitable for a society that values sustainability in economic, social and environmental terms.

In order to embed sustainable energy into the CSR framework in Japanese companies, CSR in both power producing companies and companies in general must be regarded as not only voluntary, consisting of spontaneous activities by individual companies, but also regulated by the central government. Despite the boom in CSR reporting, if each company individually sets their own standards of trustworthiness and transparency, they cannot be tracked easily. A system that relies only upon the spontaneity of companies will tend to focus on philanthropic activities that have public appeal, rather than to make efforts to enforce actual CSR progress. Voluntary rules by industrial associations and ethical norms by NGOs perform moderate roles in setting a framework, but public policy plays a significant and indispensable role in creating an overall framework for companies to readily address CSR issues, and to encourage the development of CSR in the marketplace. Thus, public policy plays several important roles in the enhancement of CSR. Firstly, it serves to create a legal framework to control business, for instance in order to protect consumers through the regulation of toxic substances. Secondly, public policy serves to create institutionalized fiscal frameworks, which serve to encourage corporations to enhance CSR. Thirdly, public policy serves to create collaborations between government, business and NGOs on CSR issues. Consequently, a "deep", systemic perspective is needed to embed CSR in Japan, which involves policy tasks at both the macro level of the Japanese government, the intermediate level of collaboration between the government and other CSR actors such as industrial associations, labor unions, consumer groups, NGOs and research institutes, and at the micro-level of the individual corporation.

Consequently, the Fukushima power plant meltdown should encourage the Japanese government to use all policy instruments to promote the energy transition; both regulation, market-based instruments and voluntary agreements made in collaboration between CSR actors and in individual companies. CSR has been a buzzword in Japan since around 2000, and the institutionalization of CSR in management has rapidly developed in most listed corporations. Many companies, including TEPCO and other electric power companies, have established a CSR department and also created the position of a CSR executive officer. But CSR management requires more than just setting up CSR management institutions; they must have appropriate authority to facilitate the practice of responsible business. Thus, CSR should be integrated into core management processes. The business of TEPCO and other electric companies is not just to make a profit, but, as TEPCO's mission statement says "to deliver electricity in a safe and stable manner". Thus, TEPCO and other electric companies are crucial players in the energy transition, and their CSR plans and practices are of huge public interest. Consequently, it is of public interest that TEPCO actually incorporates CSR into its business and management plans and practices as well as disclose non-financial data about environmental issues. Therefore, the electric companies should be made accountable for their CSR plans and practices by public policy that enforces actual CSR progress and energy transition.

References

Bendell, J. & M. Bendell. 2007. Facing Corporate Power. In Cheney et al (eds). *The Debate Over Corporate Social Responsibility*. Oxford, Oxford University Press

Blowfield, M. 2013. Business and Sustainability. Oxford: Oxford University Press

Bullis, C. & F. Ie. 2007. Corporate Environmentalism. In Cheney et al (eds). *The Debate Over Corporate Social Responsibility*. Oxford, Oxford University Press

Callenbach, E., F. Capra, L. Goldman, R. Lutz & S. Marburg. 1993. EcoManagement. The Elmwood guide to ecological auditing and sustainable business. San Francisco. Berrett-Koehler

Castells, M. 2000. Informations technology and global capitalism. In Hutton & Giddens (eds). *Global Capitalism*, New York: New Press

Chapple, W. & J. Moon. 2005. Corporate social responsibility (CSR) in Asia. A seven-country study of CSR. Business and Society, Vol.44, No.4, pp. 415-441.

Cheney, G, J. Roper & S. May. 2007. Overview. In Cheney et al (eds). *The Debate Over Corporate Social Responsibility*. Oxford, Oxford University Press

Crane & Matten. 2010. Business ethics. Oxford. Oxford Uni. Press

Deetz, S. 1992. Democracy in an age of corporate colonization. New York, State University of New York Press.

Dunphy, D., A. Griffiths & S. Benn. 2014. Organizational Change for Corporate Sustainability. London. Routledge

Friedman, M. 1962. Capitalism and Freedom. Chicago: University of Chicago Press, 1962.

Friedman, M. 1970. The Social Responsibility of Business Is To Increase Its Profits. The New York Times Magazine, Sept. 13, No. 33, pp. 122-26.

Fujiwara, H. 2006. Japan Seismic Hazard Information Station. National Research Institute for Earth Science and Disaster Prevention, Japan.

Fukukawa, K. & J. Moon. 2004. A Japanese model of corporate social responsibility?, *Journal of Corporate Citizenship*, Vol.16, pp.45-59

Gray, J. 1998. False Dawn. The delusions of global capitalism. New York. New York Press.

Keizai Doyukai, 2004. Corporate social responsibility in Japan. Current status and future challenges and CSR survey 2003. Japan Association of Corporate Executives: Keizai Doyukai

Koh, Y. 2011, How Sato sums up governments nuclear response. In The Wallstreet Journal online, , 18 April 2011.

Kolk, A. 2008. Sustainability, accountability and corporate governance. Exploring multinationals´ reporting practices. *Business Strategy and the Environment*; Vol.17, pp.1-15

Laszlo, D. 2003. The Sustainable Company. Washington. DC. Island Press

Lewin, A., T., Sakano, C. Stephens & B. Victor. 1995. Corporate citizenships in Japan. Survey results from Japanese firms, *Journal of Business Ethics*, Vol.14, No.2, pp.83-101.

Livesay, S.M. & J. Graham. 2007. Greening of Corporations? Eco-talk and the Emerging Social Imaginary of Sustainable Development. In Cheney et al (eds). *The Debate Over Corporate Social Responsibility*. Oxford, Oxford University Press

METI. 2004. The new value creation economy and evolving modalities of competition. Ministry of Economi, Trade and Industry, Tokyo

Miwa, Y. 1999. CSR. Dangerous and harmful, though maybe not irrelevant. *Cornell LawReview* 84, pp.1227–1254.

Mogi, K. 2004. Two grave issues concerning the expected Tokai Earthquake. *Earth Planets Space*, Vol.56, University of Japan, Tokyo.

Murakami, K., E. Ibuki & T. Takahashi. 2005. Co-Creation Strategy. Chitekishan-sozo, Nomura Research Institute

Ongkrutraksa, W.Y. 2007. Green Marketing and Advertising. In Cheney et al (eds). *The Debate Over Corporate Social Responsibility*. Oxford, Oxford University Press

Paine, L.S. 2000. *Value Shifts*. Harvard Business School

Pearce, J.M. 2012. Limitations of Nuclear Power as a Sustainable Energy Source. Sustainability 2012, Vol.4, pp.1173-1187

Sovacool, B. 2011. Contesting the Future of Nuclear Power: A Critical Global Assessment of Atomic Energy. Hackensack, *World Scientific*

Sovacool, B. 2011. Contesting the Future of Nuclear Power. A Critical Global Assessment of Atomic Energy. *World Scientific*. Hackensack, NJ, USA, 2011.

Sovacool, B. & B. Brossmann. 2013. Fantastic Futures and Three American Energy Transitions, *Science as Culture*, Vol.22, No.2, pp.204-212

TEPCO, 2010. Sustainability Report 2010. Tokyo: Tokyo Electric Power Company, Inc

Tanimoto, K. 2010. Structural change in corporate society and CSR in Japan. In Fukukawa (ed) Corporate Social Responsibility in Asia. Oxon: Routledge

Tanimoto, K. 2004. Changes in the Market Society and Corporate Social Responsibility. *Asia Business & Management*, Vol.3, No.2

UN. 2012. International Year of Sustainable Energy for All, 2012. Report of the Secretary General

UN. 1987. Our Common Future. Report of the World Commission on Environment and Development

Wokutch, R. 1990. Corporate social responsibility Japanese style. *Academy of Management Executive*, Vol.4, No.2, pp.56-74

Wokutch, R. & J. Shepard. 1999. The maturing of the Japanese economy. Corporate social responsibility implications. *Business Ethics Quarterly*, Vol.9, No.3, pp.527-540

Worthington, I. 2013. Greening Business. Oxford: Oxford Uni. Press

Worthington & Britton. 2009. The Business Environment. Essex: Pearson

Zorn, T.E. & E. Collins. 2007. Is Sustainability Sustainable? Corporate Social Responsibility, Sustainable Business and Management Fashion. In Cheney et al (eds). *The Debate Over Corporate Social Responsibility*. Oxford, Oxford University Press

Grassroots Denuclearization: Can Japan Denuclearize by Adopting a Renewable Energy Future?

Tony Boys and Richard Wilcox

Introduction

Following the Northeast Japan earthquake and tsunami on March 3, 2011, and with the nuclear catastrophe at Fukushima Daiichi Nuclear Power Station still ongoing, Japan has showed an increased interest in photovoltaic (PV) and other renewable energy technologies. This has been accompanied by the failure of mass demonstrations and the democratic process to lead to a denuclearization of Japan. PV panels may be seen as a possible expression of resistance to nuclear reactor restarts and nuclear power in general; an outlet for personal emotion now that the democratic process has failed to bring about the desired result, but with little realism concerning a future energy vision.

The current energy policy situation in Japan could be described as backsliding with collective heads in the sand. Japan's business/political/bureaucratic world is past-oriented, wanting to repeat previous successes but finding itself simply repeating past mistakes as it attempts to rectify the mess. This is sometimes referred to in the Japanese context as "act first think later" (*genba uchiawase shugi*), which implies a lack of planning and foresight concerning possible future problems. What is required is that Japan's energy industry and the policy-making arrangements that oversee it transform themselves into a forward-oriented force for devising an Integrated Framework for a nuclear-free energy shift that will lead to an Integrated Program of introducing decentralized renewable energy systems while slowly phasing out the old centralized systems.

Can the act of installing and using home rooftop photovoltaic (PV) panels be seen as an expression of resistance against nuclear power and, more abstractly, against the arrogance of "Jurassic power" companies and centralized electricity? Our interview survey set out to discover what ordinary Japanese people who have installed PV panels on their roofs of their homes are really thinking. At the same time, we have to recognize that electricity covers only about 22-23 percent of the total energy mix[1] in Japan and that PV panels are only one type of renewable energy technology. Home-installed PV panels, however, represent the largest contact by ordinary Japanese citizens with renewable energy, and our survey was based on the belief that the opinions of PV panel owners would be the most representative of what ordinary Japanese citizens think about renewable energy technologies and their reasons for wanting to get involved with them individually and personally.

Interviews

a) Background

Since the earthquake and tsunami disaster that struck northeast Japan on 11 March 2011, sales of PV panels have risen sharply in Japan. A feed-in tariff (FIT) scheme for renewables, under which excess energy generated may be fed back into the electrical grid at a preferential rate (expressed in terms of yen/kWh), was launched in Japan in July 2012, the total power capacity of newly operating renewable energy generation facilities reaching 3,666 megawatts (MW, 1 million watts) at the end of the first year of operation. Renewable power generation capacity increased by more than 15 percent in one year, PV systems accounting for about 95 percent of newly introduced systems. About 80 percent of the total amount of PV systems installed in past years in Japan was for residential use.

The generation capacity of newly installed PV systems one year after the introduction of the FIT scheme in 2012 was 1,379 MW for residential use and 2,120 MW for non-residential use. Thus not only has installation of rooftop PV systems increased rapidly after 2011, public and industrial use, including small scale PV arrays from around 100 kW upward and mega-solar power plants with an output exceeding 1 MW, have also seen an upward surge (Edahiro J, 2013). For comparison, a medium-sized nuclear power station has a generating capacity of around 1,000 MW.

In a joint venture with Mitsui & Co. in September 2013, CEO of SoftBank (a large Japanese telecommunications and Internet corporation), Masayoshi Son, initiated the construction of a roughly 111 MW mega-solar power plant near Tomakomai, Hokkaido, due to be completed at the end of 2015. Son was in the forefront of a number of well-known Japanese personalities who very vociferously campaigned for the end of nuclear power and a rapid move to develop large-scale renewable power generation in Japan soon after March 11, 2011 (Mitsui & Co., 2013).

According to Yano Research:

> Size of the domestic solar photovoltaic system market in FY2012 exceeded a trillion yen for the first time to achieve 1,319.8 billion yen, 180.9 percent of the size in the previous year (based on the end-user sales amount). The market is consisted of the solar photovoltaic systems for residential use that attained 704.6 billion yen (114.2 percent of the previous year), and those for public/industrial use that reached 615.2 billion yen (remarkable 545.9 percent of the previous year) (Yano Research, 2013).

It appears that ordinary Japanese citizens have several potential reasons for investing in a rooftop PV panel system. These reasons can include, but are not limited to:

1. Using the power generated by the PV panels in the home (or elsewhere) to reduce the power bill and/or to sell surplus power to the power company for the value of the FIT;
2. Partial or possible total independence from power companies' centralized power generating systems;
3. For environmental reasons, "eco-lifestyle" and so on, e.g. to combat global warming or replace nuclear power;
4. Resilience at times of emergency when power from centralized generating systems is interrupted;
5. As an expression of opposition to nuclear power by making nuclear power plants unnecessary through personal power production.

Six interviews were conducted with local Japanese people in a small, rural city in northern Ibaraki Prefecture. The location is about 120 km from Fukushima Daiichi Nuclear Power Station (FDNPS). This city was not as badly affected by nuclear fallout after 11 March 2011 when compared with areas further south in the same prefecture or with some areas of Chiba Prefecture, closer to Tokyo. However, local residents are keenly aware of the basic facts of the nuclear catastrophe. Local feeling is also compounded by the fact that the Tokai Daini Nuclear Power Station (TDNPS) is only about 20 km from the city. If TDNPS had experienced a meltdown as a result of the 11 March earthquake and tsunami, this city may well have become a permanent ghost town, as some of the towns close to FDNPS have become.

Nevertheless, since many people in the city grow some proportion of their own food and appreciate the relative cleanliness of the environment in the city, there is a strong feeling of "dispossession" among the local people. Although perhaps not to a dangerous extent, the soil, water and air have been polluted by radioactive substances from the Fukushima nuclear catastrophe.[2] People feel that this is a major loss to their lives which is irreparable except by waiting for the radioactivity to decay away over time.

There are no decontamination efforts (whether public or private) and the possible effects on children are vaguely understood, but not discussed. The authors have heard of no one "evacuating" away from the city to avoid the radioactive pollution. There is no opportunity for application for compensation from the Tokyo Electric Power Company (TEPCO) or from the government for any loss due to radioactive contamination. People are simply quietly resigned to the fact that they will have to live with a certain amount of radioactive pollution for the foreseeable future.

b) Implementation
The interviews were conducted in late November and December 2013 at the homes or business offices of the interviewees. Each interview lasted between 30 and 90 minutes (the longer interviews being for the business people). All interviews were conducted in Japanese according to a rough question sheet, which aimed to obtain relevant technical details, as well as the main reasons for

installation of the renewable technology equipment, without leading the interviewee to make statements about any specific reasons for installing PV panels or to express preferences for particular technologies. (I.e. the kinds of questions asked were: "What were your main reasons for installing the equipment?" "Do you think more people should do this?" "Do you think this kind of thing is important for Japan's energy future?" "Do you see what you are doing as in some way cooperating with the Japanese government's policy regarding Japan's energy future?" and so on). The interviewer, TB, also avoided making his personal opinion regarding specific energy technologies or policies clear to the interviewee. It was felt overall that interviewees expressed their feelings and opinions in a frank and honest manner and that answers to questions and statements made by interviewees were a valid representation of their true feelings and opinions. The authors would like to express their gratitude to everyone who cooperated with the interview survey.

Table A. Technical Details of Installed PV Panels

No.	Type	Rating	Installed	Payback	FIT
1	Business Office	6.5 kW	Nov. 2011	10 years	38 yen/kWh
	PV array	100 kW	Jul. 2013	<10 years	42 yen/kWh
2	Private home	4.18 kW	Dec. 2012	10 years	42 yen/kWh
3	Private home	4.36 kW	2008	'Never'	38 yen/kWh
4	Private home	2.46 kW	Feb. 2012	15 years	42 yen/kWh
5	Private home	3.62 kW	2006	10-15 years	48 yen/kWh
Outline of BDF Manufacture					
No.	Type	Rating	Installed		Selling price
6	BDF business	100 l/day	2008	-	105-110 yen/l

* Interviewee 1 owns both rooftop PV panels and a PV array. BDF is biodiesel fuel made from waste vegetable oil

c) Further details concerning roof PV panel equipment

None of the roof PV panel systems was accompanied by a power storage system, such as batteries. All homes and offices (1, 2, 3, 4, 5) were therefore dependent on the centralized power grid at night or on cloudy days when insufficient power was being produced to operate all electrical equipment in use in the building. In the case of interrupted supply from the centralized grid (e.g. a blackout), one electrical socket in the building was available for use during daytime and interviewees 4 and 6 expressed satisfaction regarding the fact that they had been able to make use of the power generated by the PV panels to use an electrical rice cooker, washing machine, and other appliances during the two-day blackout following the earthquake on 11 March 2011.

All interviewees stated that the reasons for not installing electricity storage equipment was that in their opinion (in some cases directly received from the PV panel installing companies) the (lead-acid) battery storage systems were both too expensive and bothersome to maintain, as well as requiring expensive replacement after several years of use. All PV panel users said they thought the power storage system technology was way behind the times and was a priority area for R&D. To different degrees, all PV panel users expressed an interest in buying power storage equipment once the price, technological capability and user-friendliness of the system reached "suitably effective" levels in terms of both energy efficiency and cost effectiveness. One home PV panel user expressed strong opinions about cutting household power bills to zero *and* holding down petrol purchases to zero or close to zero by combining the roof PV panel system with a hybrid or electric car, but admitted that this was not possible without a robust and independent power storage system.

It should also be pointed out that all private homes were "all-electric" homes, meaning that the householders use no other energy source in their homes besides electricity (e.g. do not use kerosene heaters or gas cookers). One householder installed an "Eco-cute" system, which heats water using cheap night-time electricity rates, in combination with the PV panels.

There is, in a sense, the feeling that the PV panel owners have been somewhat "cheated" by the system. The PV system has only marginally improved their resilience against blackouts and disasters – allowing the use of one electrical socket during the day only. In a winter disaster situation, heat, hot water and the ability to cook food would be crucial, but these would only be partially provided by the panels (an electric cooker would probably not be useable due to power requirement, and only one of either space heating or water heating would be available at any particular time, and then only during the day). They cannot be said to have achieved independence from the centralized power system since they rely on it at night and also rely on the power company (TEPCO) to accept and pay for the surplus energy the panels are generating. If payback times and lifetimes of the PV panels are roughly equal, then effectively all the owners have done is pay their electricity bills ten or fifteen years in advance, but to the PV panel manufacturers and installers. However, on the plus side, they have received a 15 percent subsidy on the panel equipment and also receive the roughly 40 yen/kWh FIT on the surplus energy generated by the panels during the day, which helps to shorten the payback time. Since domestic users are paying around 28 yen/kWh for electricity from TEPCO, the FIT acts as a reasonably good incentive, but the FIT rate will be reduced in coming years.

Further on the positive side for panels and other renewable power generating equipment, we can say that the cost of the equipment "*banks*" electric power against future use. Although only available at a certain maximum rate and during daylight hours, the panels do "insure" electricity for the coming 15 to 20 years, which would be very useful indeed if the centralized power system were, for some reason, to collapse five years hence. In that case, it is quite certain that some technically-minded householders would get hold of some car batteries and somehow find a way to hitch them up to their panels to provide light and other services at night.

d) Reasons for installing the PV panel equipment

One householder did not mention the environment as a reason for installing PV panels, but did express doubts about nuclear power. Three interviewees, including both the business-related respondents, expressed concerns about nuclear power as a major factor in the purchase of PV panels or operation of the BDF business. Only one householder, mentioned above, gave personal independence from the centralized grid as a major reason for installing PV panels, but did not specifically refer to nuclear power plants. These responses are not thought to be surprising given the time since the 11 March 2011 disaster and the distance of the city from the ongoing crisis at FDNPS. Although financial reasons appear to take priority over other factors, one householder emphasized personal resilience in disaster situations and did not mention financial reasons. (There were no reasons given for installing PV panels that are not mentioned here.)

Table B. Main reasons for installation of the equipment

Reason	No. of interviewees
As a business venture or for personal finances, etc.	5
Personal independence from the centralized grid	1
For the environment, "eco-lifestyle," etc.	5
Personal resilience in disaster situation, blackout, etc.	4
As an alternative to nuclear power	3

Notes: Results from all six interviews including the BDF business.
Multiple responses counted.

Discussion

Is it possible to discern from the results of this interview survey what kind of future energy policy or energy vision ordinary Japanese citizens might have in mind, and what kind of mistakes the current Abe government might be making? Additionally, what would a comprehensive and integrated long-term future energy policy for Japan look like and how might it be achieved? An attempt will be made to answer these two questions in the discussion below with reference to statements made by interviewees during the interviews.

It is not surprising that 1) there is unhappiness with nuclear power. Although only three interviewees mentioned this specifically, the idea that nuclear accidents are unacceptable is something that is so obvious in the city that it does not need to be expressed. 2) Only one interviewee (the owner of the PV panel business) had any interest in, or knowledge about, current Japanese government energy policy. This person did know, from news reports, what the rough content of Japan's new basic energy plan (finally approved by the cabinet in April 2014) was likely to be. Briefly, the energy plan at that point in time was reported to consist of the old formula of nuclear power as the base-load, thermal as the main provider (with a few provisos in order to suppress CO_2 emissions), and renewables increasing into the future, but with no specific targets for renewable energy given. However, this cannot be thought of as anything but a short-term stopgap measure. It is certainly not the comprehensive and integrated long-term future energy policy for Japan that the interviewees appear to be interested in. The reasons are, briefly:

1) There is a world uranium fuel "pinch" (Cohen and McKillop, 2012, pp.120-122; Zittel W et al., p.125-127). Even if a certain number of Japan's currently viable 40 nuclear reactors (54 pre-11 March 2011, minus 6 at FDNPS, 4 at Fukushima Daini NPS, 1 at Tokai NPS, and 3 at Onagawa NPS = 14 that are unlikely ever to be operated again), say 30 if older and relatively unsafe (e.g. for seismic reasons) reactors are discounted, how long will it be practically possible to run them given the uranium fuel pinch?
2) As mentioned by interviewees, Japan already has a crisis situation on its hands with nuclear waste. The general public knows what the problem is and wants it solved. Restarting nuclear power plants can only make the situation worse, and thus restarts can only be a temporary measure.
3) Fossil fuels are also a limited resource. It is quite possible that Japan will not be able to import sufficient oil, coal or natural gas to keep its fleet of thermal power stations (or vehicle fleet) running in 20 years' time (Brown, 2008).

Frankly, Japan has no realistic, comprehensive and integrated long-term future energy plan. This is because Japan (the government and bureaucracy, and to a certain extent business circles) has no forward-oriented long-term energy vision, effectively signifying that it has no forward-oriented long-term social vision besides "business as usual." As we have seen, this is unviable in the mid-term future, up to the 2030s, for instance. Given the current and foreseeable energy resource situation, what proposals can we make for Japan's energy policy and vision that would be practical and have a high probability of bringing about a more energy-independent Japan, a Japanese-style low-energy society, in 20 to 30 years' time?

There is a feeling that since the events of March 11, 2011 Japan could be right on the cusp of a realistic energy shift and denuclearization, which if executed properly could bring Japan out of the failing conventional, centralized-power-production-only paradigm and up close to becoming a top-class global actor in terms of innovative energy systems. What can we point to that is occurring in Japan right now and that might be the seed for such an energy shift?

Table C. Statements by interviewees on Japan's energy policy (mostly in response to the questions "Do you think this kind of thing is important for Japan's energy future?" and "Do you see what you are doing as in some way cooperating with the Japanese government's policy regarding Japan's energy future?")

Statement	No. of interviewees
Yes, I feel that what I'm doing is cooperating with government policy.	1
No, I don't feel that what I'm doing is cooperating with government policy.	3
Other statements	

1) (Private householder) Nuclear waste is a big problem that is hard to solve because there's nowhere to keep it. Fossil fuels won't be around forever, so that's why we need PV panels and other renewable energy technologies.
2) (BDF business manager) Despite the fact that nuclear power plants are now not operating in Japan, BDF does not have a future. There's a contradiction here because the government seems to want to run the nuclear power plants which are stopped now, but won't push BDF, and instead is promoting the "clean diesel" technology (to reduce CO_2 emissions). BDF, which should be an ecological fuel for the future, is therefore being discriminated against and pushed out of the market when it should be good for the environment and for future use, so how can we make the future energy transition like this? Nuclear power plants are also not good because the waste is very difficult to handle.

An Overview of Renewable Energy in Japan

a) Wind power

Wind power is one of the renewable energy sources people instantly think of along with PV panels. Global cumulative wind power capacity was 282.4 gigawatts (GW, a billion watts) in 2012, representing just over 5 percent of total world power generating capacity, up from 17.4 GW in 2000 (Wind power by country). Wind power generation is considered to be cost-competitive with other renewable energy options, and since its cost advantage facilitates large-scale introduction, it is expected to be a major factor in the expansion of renewable energy use. The number of installed wind turbines in Japan reached 1,870 in FY2011, with an installed capacity of 2.55 MW (Edahiro J. 2013). Capacity rose to 2.66 MW in 2013. The growth rate is good, around 4.5 percent, but not as high as for PV (Wind power by country). It is hard to imagine private individuals getting involved in wind power generation. The vast majority of wind power efforts currently being made in Japan are for wind farms set up by municipalities or other organizations.

One well-known example is the first municipally-operated wind farm, set up in 1989 by Suttsu Town in Hokkaido (Wind Power Stations in Suttsu Town, Hokkaido, Japan). The installed capacity is 14.55 MW from seven wind turbines (Suttu Town Office). Another early and well-known municipal wind farm is that in Shonai Town, Yamagata Prefecture (Nemoto, S., 2013). Installed capacity is 3.3 MW from six turbines (EcoPower Co., Ltd.). The former is on the west coast of Hokkaido and the latter further south on the west coast of northeast Japan, the areas considered to be the most suitable for wind power in Japan. A recent effort is the proposed floating wind farm, the Fukushima Floating Offshore Wind Farm, or Fukushima Forward, about 20 km off the Fukushima coast. One experimental 2,000-kW wind turbine began generating power in November 2013. The consortium, consisting of Tokyo University and ten companies, and which is also funded by the Ministry of Economy, Trade and Industry, plans to install two 7,000-kw turbines by March 2015. The wind farm will eventually have a capacity of 1 GW generated from 143 wind turbines (Floating wind farm debuts off Fukushima).

b) Geothermal power

Despite Japan being the world's third richest country in terms of geothermal resources, the resource has remained under-utilized. Current geothermal capacity stands at 17 facilities with a total installed capacity of 520 MW. Since most of the areas with abundant resources are located in national parks, and places where hot spring facilities already exist, making it difficult to reach agreements with hot springs proprietors, geothermal developments have been stagnant. Japan's geothermal resources are thought to be approximately 23.5 GW, 79 percent of that being within national park boundaries (Edahiro J. 2013).

A new 5 MW geothermal plant is due to start operation in Kokonoe Town, Oita Prefecture in the spring of 2015. Kyushu Electric Power Company, the owner of the new plant, currently operates five geothermal plants, three of which are in the same Kokonoe Town (Smart Japan, 2013). There is clearly room for expansion of relatively large-scale geothermal power in Japan if the current restrictions can be overcome.

c) Biomass

According to the Japanese Biomass White Paper 2013, issued in May 2013, there are currently plans to build woody-biomass power plants, many of which are large-scale plants with capacities exceeding 10 MW, at 60 or more locations nationwide. The plants will use large quantities of previously underused forest resources such as forest thinnings and other unused wood material (Matsubara H, 2013b). This is crucial for Japan's forests. Over the past 50 years of importing cheap overseas timber, the selling price of local Japanese timber has made it impossible to manage forest plantations properly. For several years after planting, it is important to thin out the young trees (and cut away excess underbrush) in order to allow the best trees to grow to their full potential, thus creating a forest that is in good condition and economically productive. The inability to do this has resulted in the current parlous state of Japanese forests.

Each large-scale power plant will require something over 100,000 m^3 of unused wood material as fuel per year. At average Japanese forest growth rates and wood density, this will require the sourcing of fuel wood from around 500 ha (2.24 km^2) of forest (Boys, A.F.F. 2000 p.102). Since Japan has around 253,000 km^2 of forest (67 percent of total land area), this does not represent a large problem (List of countries by forest area). Plants simply need to be sited in locations suitably close to both resource and consumers. Utilization of local forest resources for combined heat and power (CHP) will therefore help generate income to move forests toward the sustainable management practices that are sorely needed.

Biomass generation in Japan has in recent years used waste materials such as municipal solid waste and construction debris. However, the purchase price of power in the FIT system introduced in Japan in July 2012 was set according to the power generation method and type of fuel used. Compared to other waste-material fuels, biogas power generation via methane fermentation and woody-biomass power generation using unused wood material receive preferential treatment and have a higher purchase price. Preferential treatment for highly energy-efficient CHP has not yet been considered, and measures have not been taken to promote the dissemination of utilization of heat generated from relatively small-scale woody-biomass operations (Matsubara H, 2013b). Hopefully, this will be resolved in the near future in order to provide incentives for the planned biomass power plants mentioned above.

d) Community power revolution
Unlike rooftop PV panels, however, it is not easy for private individuals to become involved in wind, geothermal, or biomass power generation, which are generally carried out by municipalities, electric power companies or other organizations, consortia, universities or NPOs. What movements are there in Japan that might help concerned private citizens get involved in denuclearization activities through the introduction of renewable energy projects in their local areas? On March 9, 2013, an open meeting of the "Community Power Initiative" was held, in Yokohama in an attempt to kick-start a community-based energy society (Matsubara H. 2013a). An article describing the meeting states:

> Against the backdrop of the "community power revolution," a growing movement aimed at shifting energy systems from old, large-scale centralization to regionally distributed networks, the meeting focused on achievement of a decentralized network-based energy society that makes denuclearization and community independence possible by connecting local groups and key people, mutually cooperating and learning together. (Matsubara H. 2013a)

The Initiative held an inaugural meeting on June 19 and appears to be acting as an umbrella group for a number of alternative energy organizations with Institute for Sustainable Energy Policies (ISEP) at the core. Perhaps it will provide a vehicle for larger numbers of private individuals to become involved in local renewable energy projects.

ISEP's Japan in 2050 Vision

In 2008, the Institute for Sustainable Energy Policies published a "2050 Renewable Energy Vision," which is a full energy scenario for Japan in 2050. For the purposes of denuclearization, it is appropriate to refer to the power supply section of this document. The following table is reproduced from p.5 of the document (ISEP), with the two right-hand columns added by the authors for reference.

Table D. Japan's Power Supply in 2050

Type	Grid power (TW)	Decentralized power (TW)	Total power (TW)	Installed capacity (GW)	Installed capacity 2011 (GW)	Renewable capacity 2011 (GW)
Coal	30.0	11.5	41.5	5.6	38.77	
Oil	0	0	0	0	46.55	
Gas	128.9	35.3	164.2	34.2	65.53	
Nuclear	64.4	0	64.4	11.0**	48.96	
Hydropower (pumped)	8.7	0	8.7	19.8	26.24	
Hydropower	118.1	1.3	119.4	27.6	20.76	
Geothermal	72.0	10.6	82.6	11.8		0.53
Biomass	35.9	82.3	118.2	15.9	5.7	2.50*
PV	15.0	135.0	150.0	142.7		3.00
Wind power	87.6	0	87.6	50.0		2.55
Total	560.5	276.1	836.6	318.5	245.38	8.58
Renewable %	59%	79%	67%	78%	2.3%	3.5%

Notes: Power supply capacity for 2011 is from: http://www.enecho.meti.go.jp/topics/hakusho/2012energyhtml/2-1-4.html, Graph No. 214-1-5, Trends in Installed Power Capacity (General Power Business). However, METI estimates renewables at 5.7 GW, which seems low considering the figures already given in this paper. The figures already citied above are therefore reproduced in column 7 with an estimate for biomass. * Estimated from http://www.asiabiomass.jp/topics/1212_03.html
** The Tokai Daini Nuclear Power Station reactor is rated at 1.1 GW, so this figure of 11 GW represents roughly the capacity of 10 medium-sized nuclear power reactors.

What is obvious from the table is that under the 2050 Vision scenario Japan does not plan to denuclearize, even by 2050. ISEP's position at the time was that nuclear power can be operated safely and should be part of the energy mix to 2050 and perhaps for at least a decade later. Recent opinion polls consistently show that this is not what Japanese people really want. (E.g. in response to Question 14 of a nationwide interview opinion survey three years after the earthquake disaster, 68.9 percent of respondents were in favor of a nuclear phase-out (Tokyo Shimbun, March 9, 2014, p.2).

Oil is reckoned at zero in 2050, where gas is predicted to halve, and hydropower (though not pumped-storage hydropower) is predicted to increase somewhat. In renewables, PV appears to be the big "winner" with wind power trailing a long way behind. Biomass, at just under 16 GW, appears to be still underutilized. Geothermal power, as we have seen above, is reckoned to achieve about half its potential.

Overall, installed capacity rises to 318.5 from 245.38 in 2011, which appears to promise a more "affluent" lifestyle than at present, but actually does not because the capacity utilization factor will drop due to the intermittent nature of PV and wind power. That is, PV and wind will not operate 24 hours a day because they are dependent on the intermittent nature of the sun or wind, unlike geothermal or biomass power plants, which could theoretically operate 24 hours a day, giving a higher capacity factor and therefore generating power close to the capacity rating × time calculation. It's very reassuring to think that this much power might be available in 2050, but realistically we have no idea of how much power society will require in 2050. Assuming that the general trend in social values is toward a more sustainable low-energy (or low-carbon) future economy/society that is far less dependent on nuclear power, it would appear reasonable to forego the 11 GW of nuclear power and for Japan to distance itself as far from the Jurassic power paradigm as possible by 2050. That would seem to be more like what ordinary Japanese people want as a future energy plan, and in terms of reduced additions to the mountains of nuclear waste, but are they likely to get it?

Two recent Japanese government policies suggest they won't. One is the ending of the national residential PV subsidy program (Movellan, J. 2014). Most of the households interviewed above received a 10 percent national (and 5 percent local) subsidy on the price of their panels, worth about USD3,000 if the panels cost 3 million yen (USD30,000). This is a considerable disincentive to future panel buyers.

The second is that Japan's Basic Energy Plan was approved by the cabinet on April 11, 2014, officially abandoning the zero-nuclear goal of the previous Democratic Party of Japan administration by pledging to encourage restarts of Japan's currently idled nuclear reactors. The approved energy plan also failed to set specific numerical targets for a desirable ratio of energy resources for Japan, including oil, gas, nuclear power and renewable energy (Yoshida, R).

Shinzo Abe's LDP government is clearly intent on moving in the opposite direction to that hoped for by a majority of Japanese people.

Can Japan Learn Something from Germany's *Energiewende*?

"12 Insights on Germany's *Energiewende*" (12 Insights on Germany's Energiewende, 2013) is an extremely valuable source for the insights it can bring to our understanding of the current Japanese energy situation and the direction it might take in the future.

Firstly, wind power and PV emerged as winners from the technology competition initiated by the German Renewable Energy Act. They were identified as the most cost-effective technologies with the greatest potential in the foreseeable future. The report notes that "all other renewable technologies are either significantly more expensive or have limited potential for further expansion (water, biomass/biogas, geothermal energy) and/or are still in the research stage (wave power, energy from osmosis processes, etc.)" (p.5). Having expertise in these technologies, it would seem reasonable that Japan should push as hard as possible to expand the installation of these where appropriate.

The report also informs us that "Wind and solar power have operating costs close to zero" (p.22). Once they are installed, the marginal cost of generating a further kWh is effectively zero. This is quite different from conventional power stations, especially nuclear ones. Another feature of this technology is that "Wind and PV power should be expanded in tandem since they have mutually complementary features; generally speaking, the wind blows when the sun is not shining and vice versa" (p.8). Thus, as far as reasonable, given local resource endowments, PV and wind should be developed together in roughly equal amounts.

Interestingly, the report states that "Wind and PV will form the basis of the power supply, with the rest of the power system being optimized around them; most fossil-fueled power plants will be needed only at those times when there is little sun and wind, they will run less hours, and thus their total production will fall: 'Base load' power plants will be a thing of the past" (p.9). Despite the insistence of Japan's new Basic Energy Plan, nuclear power plants are therefore not necessary, since the base load is no longer "important." Japan should thus carry on doing exactly what it is doing now – let the nuclear plants stand idle – and then begin procedures for decommissioning them.

Naturally, "…fluctuations in wind and PV production will demand significantly greater flexibility from the power system." (p.11), but "Technical solutions to provide sufficient flexibility readily exist today" (p.11). "The challenge is not about technology and control, but rather about incentives" (p.13). Disincentives for renewables, as mentioned above, are therefore a move in the wrong direction. The answer would be to shift incentives (subsidies) now provided to the nuclear power industry (including all expenditures for the nuclear fuel cycle) to renewables such as PV, wind power and CHP systems based on forest biomass. The Japanese government budget for nuclear administration, research and subsidies awarded to municipalities that host nuclear power stations amounted to about 389 billion yen in 2012. This was down about 10 percent from 2011, but an additional sum of just under 500 billion was earmarked for "Projects Associated with the TEPCO Fukushima Daiinchi Nuclear Power Station Accident" (Citizens' Yearbook of Nuclear Power, 2013: pp. 302-305). Denuclearization would therefore immediately make available about 400 billion yen of government money for the promotion of nationwide renewables projects.

The report goes on to state that "The flexibility options … are already technically available today and can be implemented at relatively low cost. Since they primarily involve "large-scale" facilities (CHP plants, biomass facilities, industrial processes, large heat storage), the problem of controlling them is technically easy to solve..." (p.13). So the idea of using Japan's underutilized forest resources to harvest fuel wood sustainably for CHP facilities fits in precisely with the idea of expanding PV and wind power. Without the need to provide subsidies for a nuclear program, Japan could afford to help municipalities and other local organizations set up PV and wind power systems backed up by CHP systems.

As noted above, the householders possessing PV panels regretted the fact that they were unable to store the power created during the day for use at night or for recharging an electric car battery. The report notes, however, that "Grids are cheaper than storage facilities: Grids decrease the need for flexibility: fluctuations in generation (wind and PV) and demand are equilibrated across large distances" (p.14). Thus local (possibly "smart") grids based on PV, wind power, local hydropower and local CHP from forest resources can solve this power storage problem. Moreover, with municipal-sized grids, solving the problem across "large distances" is not necessarily required.

Finally, the report notes that "Energy efficiency decreases total costs; increased energy productivity enables the decoupling of economic growth from energy consumption" (p.31). Thus there is a need to give serious consideration to implementing a gradual shift to a low-energy society rather than to assume that the same or more power will be "needed" in, say, 2050 as we "need" now. But this will require creative transitions on the part of all actors in society, businessmen, politicians, bureaucrats, academics and researchers, private individuals and others instead of the slavish subjugation to a system that cannot foresee, or will not permit, change.

We may be in a position now to state what ordinary Japanese people would "like" to see as an **Integrated Framework.** That would lead to the introduction of an **Integrated Program** for the energy transition to a future low-energy society. An integrated framework would be firstly flexible and secondly non-exclusive.

We can quite easily see that rural communities and residential areas in cities could be encouraged to go over to *totally decentralized, small scale, short distance, low-voltage, low-power electric power distribution networks (local grids) based on local endowments, but primarily using PV power, wind power, CHP forest biomass use, local hydropower, geothermal power and so on in proportions that match with local endowments.* This could be backed up by a connection to the national grid to provide power if and when necessary. This currently possible, flexible and non-exclusive system would be safe and resilient because it is diversified and decentralized. It would not easily be knocked out by storms, earthquakes or other disasters, and even though some parts might be damaged they would be easily replaced, bypassed, repaired or retired while the community suffers little impact from loss of power supplies. Since such a system would be ideal for rural areas, local forest resources would have an important part to play, as mentioned above. The system can, and should, also be linked to a strong project for local BDF production. BDF can be used to fuel crucial farm machinery should there ever be a diesel fuel shortage, and can also be used to operate diesel power generators in emergencies.

That goes a long way to solving part of the problem, but what about heavy industrial power users? Note that the ISEP 2050 Vision above still allows for about 25 percent of power (even after eliminating nuclear power) to be generated from conventional sources and a part of the power generated from renewables can also be fed into the national grid. So it is not as if a quite remarkable shift towards renewables in 2050 means eliminating industry, but rather that industry will have to become leaner, more efficient, more creative, more geared to a low-energy society. It would also mean, of course, that before very long society will have to start to consider what the main values of such a society would be and begin to move toward them in a forward-oriented way instead of insisting on past-oriented ideas about how to maintain or revive the high economic "Jurassic" smokestack high-speed economic growth of the 1960s and 70s. More suggestions will be given below.

Is the Japanese Nuclear Village Invincible (and Japanese Denuclearization Impossible)?

As we have seen in our earlier paper in this volume, recent events have shown that despite the Fukushima nuclear disaster Japan's "nuclear village" is still extremely robust and there appears to be little opportunity for the people of Japan to realize a nuclear power phaseout in the near future. Japanese anti-nuclear activists should perhaps take a closer look at the effects of the Fukushima disaster in France as seen through the eyes of a European energy expert.

> France is unquestionably the 'poster child' for the nuclear lobby. It gets about 78 percent of its electricity from nuclear power.

> French former state-owned monopoly Électricité de France (EDF) is the operator of all 63 French reactors, and EDF in 2012 supplied about 92 percent of all electricity consumed in France. The state still has corporate capital control and policy control of EDF, as well as Areva.

> Only since Fukushima has there been a notable shift in French political values, viewpoints on energy, and energy policy and programs. Until 2011 the backlog of pro-electricity and pro-nuclear policy was overwhelming. This has broken up.

> The pro-nuclear pro-electricity policy of France was almost certainly replicated by Japan and a large number of other developed countries. Therefore the change away from this

'monolithic policy' can only be a rupture or break, easily able to lead to or cause policy and political confusion.

Until 2011 there was no questioning in France of national, regional and local building regulations making electricity the mandatory and sole form of residential heating, for example. Inevitably this created a huge structural demand for electricity.

However, since 2011, one completely fundamental basis of French nuclear-electrical energy economic policy has gone. This was the previously total commitment to cheap electricity. From this year 2013, EDF will raise its prices by 10 percent every year for at least three years, perhaps four years. This rate of price increase is at least five times the official or admitted CPI rate and ten times the permitted annual pay rise of all government employees (1 percent a year).

French energy economic policy to encourage – or force – the consumption of electricity has been literally turned on its head. Today, many urban highways are unlit or only partly lit and all shops are legally required to turn off their window lighting by 10pm, for example, and most do it.

EDF has also been mandated or tasked by the state with operating FITs for renewable energy producers. EDF is obliged to buy power (as in Germany and most other EU countries) at inevitably highly politicized elevated tariffs further increasing EDF's extreme high corporate debt.

The previous monolithic national policy of 'all-nuclear cheap power' has disappeared in at most three years. However, what comes next is very unsure due to policy confusion or cognitive dissonance.

Possibly the biggest difference between the Japanese situation and the French is that Japan (for as long as the reactor fleet is not 'turned back on,' or powered up to net surplus power operating levels) is obliged and forced to make a **rapid transition**. The French policy, to the extent it clearly exists amid intense political debate, is for a 'gradual taper down' of nuclear power. The present outline goal of the Parti Socialiste is '25 percent less nuclear power by 2025'.
(Andrew McKillop, personal communication, December 1, 2013)

The time for that "rapid transition" is now, while all the nuclear reactors are shut down and Japan is effectively in a state of temporary denuclearization. This temporary state should be made permanent. Japan must bite the bullet and find creative solutions for the political and economic problems of re-employment for those in the nuclear industry, the decommissioning of all nuclear reactors in Japan, the current account deficit due to the increased imports of natural gas and oil for thermal power generation, the loss of export orders for nuclear reactors caused by abandoning nuclear power, and the role of propping up the myth of the "peaceful atom" that has been forced on Japan because it has civil nuclear power but apparently no nuclear weapons.

These problems, like the huge Japanese national debt, will not be solved by a wish to return to the 1960s or 70s, or even to the 90s and 2000s, when Japan had over 50 operational nuclear reactors. They will only be finally solved, and solved to the satisfaction of the majority of the people of Japan, when the decision to restart the reactors is rescinded and the government begins to lead the people forward to the sustainable low-energy society and economy of the future.

What Can be Done in this Rural City?

The rural city where the interview survey was carried out, for example, has two local rivers which have a number of floodgate-type dams used to supply irrigation water for wet rice cultivation. These could be easily adapted for small-scale hydroelectric power generation. Mountain forest also accounts for approximately 62 percent of the city area, forest resources currently being pitifully undermanaged and underutilized due to low demand for domestic timber. Sustainable forest management could harvest fuel wood, part of which could be used to fire one or more CHP plants which could pipe surplus hot water to residential areas. As the longest distance from locations where power plants might be located to the furthest point in the city beyond the main residential area is about 25 km, this is relatively easily done using insulated steel pipes buried a few meters below the surface of roads.

Table E shows energy use for a typical part-time farming household (where the main income-earner is also a part-time farmer, or farm work is performed by person(s) other than the main income-earner), of which there are many in Japan's rural areas, and how that energy might be supplied by renewable energy sources in a low-energy society.

Table E. Local energy source substitutes for current energy uses

Operation	Now provided mainly by:	Energy use per household	In future may be provided by:	Rough equivalent in alternative energy sources
Cooking food (including hot drinks)	Electricity, gas	3,517 MJ/yr[a]	Wood, thermal solar	190 kg/yr dry wood (e.g. branches)
Heating water for washing and bathing	Electricity, kerosene, gas, thermal solar	13,536 MJ/yr[a]	Wood, thermal solar, CHP	720 kg/yr dry wood (e.g. branches)
Space heating for dwelling during five winter months Nov.-Mar.	Electricity, kerosene, wood	12,389 MJ/yr[a]	Wood, CHP	660 kg/yr dry wood (e.g. branches)
Electricity for the household water supply	Electricity	Depends on water supply system, but not more than 100 kWh/yr/ household	CHP, hydropower or PV.	Easily covered by small PV panel if supplied from a household well.
Lighting	Electricity	Roughly 150 kWh/yr[j]	PV, CHP, hydropower, candles	26.3 kWh/yr[b]
Refrigeration (especially for specialized needs, e.g. hospitals)	Electricity	385.4 kWh/yr[c]	CHP, hydropower or PV	385.4 kWh/yr
Transport	Gasoline, diesel, electricity (trains)	500 l gasoline or diesel fuel[h]	BDF, bioethanol, bicycles	Reduce use to essential transport only – use public transport
Farm machinery	Gasoline, diesel	110-140 l/ha/yr[d]	BDF, bioethanol, farm animals	Approx 100 l/ha/yr BDF or bioethanol[e]
Electricity for farming operations: irrigation, post-harvest drying, milling and processing	Electricity	30 kWh/ha/yr[f]	CHP, hydropower or PV. Sun-drying, manual power, with or without mechanization, e.g. hand-powered winnowing machines, etc.	Perhaps possible with a household 3-4 kW PV panel system[g]

Notes:

a. The Energy Conservation Center, Japan, Handbook of Energy & Economic Statistics in Japan, 2007, p.93. See Table 5 below for calorific value of wood. MJ is megajoules, one million joules of heat energy.

b. Based on one 18w lamp for 4 hours per night for 365 days per year. Easily supplied by one small PV panel.

c. Based on a 300-400 liter family refrigerator running at an average power use of 44W/hr for 24 hours per day (1.056 kWh) for 365 days per year. Easily supplied from a household PV panel of 3-4 kW. Figures from the Seikatsu Club website: http://www.seikatu-cb.com/kiwami/siyou02.html

d. Williams, James H., David von Hippel, Peter Hayes, "Fuel and Famine: Rural Energy Crisis in the Democratic People's Republic of Korea," Institute on Global Conflict and Cooperation, IGCC Policy Papers, Paper #46, March 2000, (http://repositories.cdlib.org/igcc/PP/pp46) p.8-9

e. Assuming BDF is produced from a secondary crop of e.g. rapeseed, the oil being retrieved as WVO after use for cooking, or that ethanol can be distilled from processing of agricultural waste (provided soil fertility conditions are met) or crops that use land other than that used for human food, and that current machinery can be adjusted to run on ethanol.

f. Pimentel, D., (Ed.), *Handbook of Energy Utilization in Agriculture*, CRC Press, 1980, p.96. The figure given here is about 16 kWh/ha/yr for drying harvested rice. Roughly the same amount may be used for other processing.

g. Very heavy seasonal use, but a 3 kW panel might provide 15 or more kWh of electricity on a sunny day.

h. Assumes each car is driven around 10,000 km/yr and fuel consumption is around 20 km/l.

j. Assumes four 18 w lamps are used for 6 hours each night for 365 days per year.

Note on energy obtained from mountain forests

Energy for heating could be supplied by fuel wood from local forests either by direct burning of wood or hot water supplied by CHP. Since around 62 percent of the city area is mountain forest (216 km^2 – Land Area by Land Type, 2012), we can make a theoretical calculation of the amount of wood and energy that can be sustainably harvested per year from those forests, as in Table F below.

Table F. Calculation of energy value of sustainably harvested wood

Average annual growth per ha for Japanese forests	13.8 tonnes
Multiply by 0.4 to eliminate leaves and branches	5.52 tonnes
Multiply 21,600 ha x 5.52 tonnes for total harvested weight	119,000 tonnes
Multiply by 18.84 10^9 J/tonne to obtain energy value in J	2.25 10^{15} J

Sources: Private interview with Kouichiro Koike at Shimane University, Matsue City, 7 February 2000, and Kishi, Sadakichi, Considering Forest Energy, 1980, pp.31, 32, 59

The calculation is carried out as follows.

Convert the household energy use, about 11 million kcal/year, to MJ.

11,000,000 kcal/household × 4.1868 10^{-3} MJ = 46,000 MJ/household (Source as in 'a', Table E)

Calculate the energy use by all households in the city:

× 16,000 households = 7.36 10^8 MJ = 7.36 10^{14} J

Calculate what percent of the sustainably harvested forest wood would supply this energy:

7.36 10^{14} J / 2.25 10^{15} J × 100 = 33% (approximately 2.45 tonnes/household)

(Or 65 percent of this if use is limited to the amounts of energy mentioned in the first three items of Table E)

Thus 22 percent to 33 percent of sustainably harvested wood from city forests would theoretically cover all current household energy needs for the city (not counting harvested branches), but this would need to be multiplied by the coefficient of efficiency, somewhere in the region of 0.5 to 0.8, in order to obtain the amount of wood fuel required to actually fulfill the purpose intended (EPA).

As mentioned above with reference to biomass use in CHP, a large CHP facility would require around 100,000 m^3 or 6,500 tonnes of fuel wood per year, but this would include branches, and therefore can be sustainably harvested from about 500 ha of forest. As the city has 216 km^2 (21,600 ha) of forest, running several large CHP facilities does not present an insurmountable problem. In fact, as mentioned above, it is more likely to be highly beneficial for the forests.

Leaves fallen on forest floors may be a very important source of fertility for farmland. Branches may be used directly as fuel for household uses or converted to charcoal for various uses, possibly including transport.

However, all of this will depend to a certain extent on:
1) The condition of the forests,
2) The (predominant) kinds of trees in the forests[3]
3) Forest ownership (private, municipal, prefectural, state, etc.)[4]

If ready cash were to be available for wood fuel or timber sustainably harvested for the purposes of new CHP plants or local construction there is very little doubt that owners would be queuing up to sell. Repeated on a nationwide scale, this would go a long way to solving Japan's notorious forest management problems.

Wood will also be needed for construction purposes and for other purposes such as furniture-making and so on. All of this will have to be carried out in an orderly and organized manner that is not destructive of forests. It *must* be done sustainably. The forests should not be destroyed or downgraded for short-term reasons, in fact the opposite. The forests should be slowly expanded and improved so as to provide maximum sustainable productivity for future generations.

It should be clear from the above that this rural city can be self-sustainable in energy from renewables through the use of forest resources, local hydropower, small-scale PV and agricultural resources (oil seed for BDF production) and can make the transition to a low-energy society by constructing several CHP facilities and laying insulated pipes throughout residential areas. This would require considerable financial resources and time (5 to 10 years might be needed to complete a project of this nature) but would effectively make the city independent of the national grid and centralized "Jurassic" power suppliers.

Conclusion

While Japan currently (May 2014) has no nuclear power stations in operation, realistic denuclearization would need an Integrated Framework for an energy transition that would lead to an Integrated Energy Transition Program. A slow phasing out of fossil fuels for thermal power generation would also be necessary to solve Japan's current account deficit problem. Such a program would include both extremes of the scale of potential responses; the individual household and the large-scale industrial complex.

1. Extreme large-scale proposals (installing local smart grids everywhere in Japan would itself be an extreme-scale project) for the continuation of the conventional industrial system, albeit with a certain amount of adjustment as consumption activities exhibit changes associated with the transition to a low-energy society and economy. This would entail, for example, less transportation due to local production for local consumption and the manufacture of fewer throw-away goods and more durable, easily-repaired goods.

2. Small-scale responses would include, incremental, ground up, reasonably low-cost projects initiated and implemented by municipalities, NGOs, neighborhood associations, and small enterprises.

For the extremely large-scale systems, besides the maintenance of a certain amount of conventional power generation, as in the ISEP 2050 Vision, but minus the nuclear power stations since the objective is to denuclearize, we propose the introduction of ocean thermal energy conversion (OTEC) systems, geothermal energy, and the possible use of deep ocean current turbines.

OTEC uses the temperature difference between cooler deep and warmer shallow or surface ocean waters to run a heat engine and produce useful work, usually in the form of electricity. OTEC is a base load electricity generation system, i.e. it delivers electricity 24 hours a day all year long. However, small temperature differentials impact the economic feasibility of ocean thermal energy for electricity generation. Water pumped up from the deep ocean is nutrient rich and therefore this technology can be associated with various kinds of fishing activities. One Japanese example began in April 2013 on Kume Island in

Okinawa Prefecture. The pipe for the OTEC had been previously used for the intake of deep sea water for fishery and agricultural use. (Ocean thermal energy conversion)

Geothermal energy has been mentioned above. However, even the ISEP scenario only envisages making use of about half of Japan's potential resources. While a large proportion of the resources are in national parks, it should be possible to devise ways to utilize them without significant environmental deterioration. As geothermal resources exist in rural areas or national parks, the electricity and surplus hot water supplied can be used for forestry and agricultural applications, including for food processing, which adds value to agricultural raw materials by turning them into marketable products.

For small-scale systems we propose district and local CHP systems, as mentioned above, with tidal and sea current energy for coastal locations, to be operated in combination with PV, wind and local hydropower and other suitable systems. These should be introduced in proportions that accord appropriately with local endowments and which are likely to provide the most balanced and stable power supply to the community. Despite misgivings about competition between human food and fuel for the use of farmland, this can be combined with BDF production (from WVO originating from oil seeds, e.g. rapeseed, grown in the local area) and bioethanol production from waste forest products as a substitute for gasoline in gasoline engines (i.e. for cars, trucks and many smaller farm machines). However the overall balance of the use of forest resources has to be managed carefully in order to ensure that sustainability is maintained.

As argued above, the implementation of such an Integrated Program (based on an Integrated Framework) could be carried through by shifting subsidies now provided for nuclear power to an integrated national program of renewable energy system introduction. Japan's existing nuclear reactors would then be slowly decommissioned, and hopefully the catastrophe at Fukushima Daiichi Nuclear Power Station can be resolved without further massive pollution or mishaps. This would require the drawing up of a viable Integrated Framework for the Integrated Program, the basis of which is mentioned above. In turn, this would necessitate that forward-oriented, constructive and creative planning take the place of Japan's current past-oriented energy policy culture with its total lack of integration and coherent thinking.

Notes

1. The 2013 Energy White Paper Overview of Japan's Energy Balance Flow (FY2011) Figure No. 211-1-2 shows the contribution of electricity to total final energy to be 22.7 percent. (By calculation: total final energy was $14,527 \times 10^{15}$J and electricity 3297×10^{15}J.) http://www.enecho.meti.go.jp/topics/hakusho/2013energyhtml/2-1-1.html.

2. The city hall announced in late March 2011 that the soil in paddy fields in the southern part of the city was 92 Bq/kg. An air dose monitoring station in the

south of the city was showing around 40 nGy/h just after midnight on March 15, 2011. At 0530 that morning the value shot up to 2241 nGy/h (probably Unit 3 at Fukushima Daiichi Nuclear Power Station exploding – see NIT 142 p.4, http://www.cnic.jp/english/newsletter/nit142/index.html), receded to 1391 nGy/h at 0630 and then rose to 2668 nGy/h at 0700. Thereafter the value declined gradually all the way down to 150 nGy/h. This was still roughly four times the 'normal' background level of about 38 nGy/h. As of May 1, 2014 the air dose level was given as steady at 46 nGy/h.

3. A large percentage of the city's forests are "Japanese cedar" (*sugi,* Cryptomeria japonica), planted in the post-war period when there was a shortage of construction wood. These forests are good for construction wood, but may be unusable for fertilizer.

4. Forest ownership: According to the 2005 Ibaraki Prefecture Agricultural and Forestry Census, p.208, the city's forest area is 21,153 ha (of a total area of 34,838 ha; 59 percent), of which 11.9 percent (2,521 ha) is state-owned, 84.6 percent (17,892 ha) is privately owned, and 3.8 percent (814 ha) is publicly owned (figures rounded).

References

Boys, A.F.F. (2000). Food and Energy in Japan: How will Japan feed itself in the 21st century? PDF Retrieved from http://www9.ocn.ne.jp/~aslan/21fee.pdf

Brown, J and "Khebab" (January 7, 2008). A Quantitative Assessment of Future Net Oil Exports by the Top Five Net Oil Exporters, retrieved from http://www.graphoilogy.com/2008/01/quantitative-assessment-of-future-net.html

Citizens' Yearbook of Nuclear Power, 2013, Citizens' Nuclear Information Centre.

Cohen M and McKillop A. (2012). The Doomsday Machine, Palgrave Macmillan.

EcoPower Co., Ltd., Corporate Profile. Retrieved from http://www.eco-power.co.jp/project.html (More Japanese wind farms are mentioned here.)

Edahiro J. (2013, November 29). Renewable Energy in Japan -- Current Trends Show Promise and Opportunities, Japan for Sustainability Newsletter #135. Retrieved from http://www.japanfs.org/en/news/archives/news_id034505.html

EPA, Combined Heat and Power Partnership, Efficiency Benefits. Retrieved from: http://www.epa.gov/chp/basic/efficiency.html

Floating wind farm debuts off Fukushima, The Japan Times, November 11, 2013, http://www.japantimes.co.jp/news/2013/11/11/national/floating-wind-farm-debuts-off-fukushima/#.UwWGCmJ_vh4

ISEP, (2008, June 3). 2050 Natural Energy Vision (Japanese), http://www.re-policy.jp/2050vision/2050vision080603.pdf

Land Area by Land Type (2012). Hitachi Omiya City (Japanese). Retrieved from http://www.city.hitachiomiya.lg.jp/page/page000205.html

List of countries by forest area, Wikipedia,
    http://en.wikipedia.org/wiki/List_of_countries_by_forest_area

Matsubara H. (2013a, June 3). Community Power Revolution Spreads in Japan,
    Japan for Sustainability,
    http://www.japanfs.org/en/news/archives/news_id032838.html

Matsubara, H. (2013b, August 1). Use of Forest Resources Increases due to Rapid
    Expansion of Biomass Generation, Japan for Sustainability,
    http://www.japanfs.org/en/news/archives/news_id032966.html

Mitsui & Co. (2013, September 24). News release, Construction of mega-solar
    power station at Abira Town, Hokkaido, Retrieved from
    http://www.mitsui.com/jp/ja/release/2013/1201036_4689.html

Movellan, J. (2014, February 20). Japan FIT Changes Reflect End of Residential PV
    Program and Delay in Non-residential Projects, renewable Energy
    World.Com, Retrieved from
    http://www.renewableenergyworld.com/rea/news/article/2014/02/japan-fit-
    changes-reflect-end-of-residential-pv-program-and-delay-in-non-residential-
    projects?cmpid=WNL-Friday-February21-2014

Nemoto, S. (2013, June 18). Farmers' 'nemesis wind' losing power to turbines,
    scientist says, The Asahi Shimbun, Retrieved from
    https://ajw.asahi.com/article/behind_news/social_affairs/AJ201306180008

Ocean thermal energy conversion, Wikipedia. Retrieved from
    http://en.wikipedia.org/wiki/Ocean_thermal_energy_conversion

Present Status and Promotion Measures for the Introduction of Renewable Energy
    in Japan. (2014). *The Ministry of Economy, Trade and Industry (METI).*
    Retrieved from
    http://www.meti.go.jp/english/policy/energy_environment/renewable/

Smart Japan, (2013 November 25). Geothermal power station on the border of Oita
    and Kumamoto, 5MW to begin operation in 2015 (Japanese)
    http://www.itmedia.co.jp/smartjapan/articles/1311/25/news018.html

Suttu Town Office, Wind Turbines. Retrieved from
    http://www.town.suttu.lg.jp/huusya/huutahuuryokuhatudensyo..html

12 Insights on Germany's Energiewende. (2013, February). *Agora Energiewende.*
    Retrieved from http://www.agora-energiewende.org/topics/the-
    energiewende/detail-view/article/12-insights-on-the-energiewende/

Wind power by country, Wikipedia,
    http://en.wikipedia.org/wiki/Wind_power_by_country

Wind Power Stations in Suttsu Town, Hokkaido, Japan,
    http://hitachi.com/environment/showcase/customer/case_vol1/index.html

Yano Research. (2013). Solar Photovoltaic System Market in Japan: Key Research
    Findings 2013 http://www.yanoresearch.com/press/press.php/001151

Yoshida, R, (2014, April 11) Cabinet OKs new energy policy, kills no-nuclear goal, Japan Times, April 11, 2014, http://www.japantimes.co.jp/news/2014/04/11/national/cabinet-oks-new-energy-policy-kills-no-nuclear-goal/#.U2Ll8oF_t8E).

Zittel W, Zerhusen J, Zerta M, Fossil; and Nuclear Fuels – The Supply Outlook, Energy Watch Group, March 2013, http://aie.org.au/AIE/Documents/EWG-update2013_long_18_03_2013.pdf

European Energy Transition - Japan's Non-Nuclear Future

Andrew McKillop

Distressed Nuclear Assets

According to the US Energy Information Administration (EIA), nuclear generation in Japan covered about 26 percent of its power supply prior to the 2011 earthquake and was (the US EIA claims) one of the country's least expensive forms of power supply. According to Turkey's Hurriyet Daily, 18 February 2014, the global share of nuclear energy in the world's power generation declined steadily from a historic peak of 17 percent in 1993 to about 10 percent in 2012, pushing down nuclear power's share of global commercial primary energy production to 4.5 percent, a level last seen in 1984. This 4.5 percent share for nuclear power in world primary energy, and its 10 percent role in world electricity supply, showed that about 45 percent of primary energy is used for power production – but electricity rarely covers more than 20 percent of final energy demand in the advanced industrial countries, and usually far below 10 percent in developing countries.

The so-called "dependence on nuclear power" that is often bandied around by politicians and of course by the nuclear lobby, can be understood as only concerning a small, or very small part of world energy. For nuclear weapons, of course, the story is different. Building nuclear weapons obligatorily needs nuclear reactors. Meeting national energy needs does not, as is shown also by the number of countries utilizing at least 1 nuclear power reactor standing at 31 as of January 2014 (according to Euronuclear), compared with the total of 192 member nation states of the U.N. as of early 2014. In other words, 161 countries, today, do not use nuclear power.

Japan replaced the loss of its nuclear power with generation from imported LNG (natural gas), low-sulfur crude oil, fuel oil and coal, and a very small but growing supply from the so-called "new renewables" (wind and solar power). The US EIA in late 2013 said fuel import cost increases have resulted in Japan's top 10 utility companies losing over $30 billion in the past two years.

Japan's total spending on fuel imports in 2012 was $250 billion, accounting for about 33 percent if its total imports by value. Japan's trade deficit for year 2013, using official data was about 11.5 trillion yen ($113 billion), close to two-times its year 2012 trade deficit.

In Europe, the March 2011 earthquake and tsunami disaster in Japan significantly increased political support to Europe's post-2008 energy transition program, especially in Germany. This program, excepting the now-reduced and totally uneconomic biofuels development program included in the European

Parliament's "20-20-20 climate-energy package" vote of December 2008, almost entirely focuses on the electric power sector. In that sector – which is the only one that concerns nuclear power – European plans as of present aim for a near-100 percent 'backing out' of fossil fuels, and their near-total replacement by renewable energy for electricity production, by about 2045.

In the German case, while EU28 member state disharmony on this issue is important and increasing, national energy policy features the total abandonment of nuclear power by 2022. This was added as an official national energy goal from 30 May, 2011, following the Fukushima disaster. At the time, in 2011, nuclear power provided about 17.7 percent of German total electricity supply.

By 2013, renewables covered about 26.5 percent of total German electricity supply. On a "prima facie technical basis", therefore, Japan can relatively easily replace all its nuclear generation within a relatively short period of time but economic, financial, industrial and technical challenges exist.

The European energy transition plan, which as already noted almost exclusively concerns power generation, has however and to date engendered many diseconomies, in major part due to the financialization of the power sector, like other economic sectors. In my opinion the diseconomies of energy transition in Europe are not primarily due to program goals, but are due to an incompatible mix of the now mainly short-term private corporate aims and ambitions of the "legacy power producers", and longer term national goals regarding energy policy. Due to power sector financialization and regulatory change, and also due to "legacy technology" issues, this results in major uncertainty.

Notably in Germany, but also in other EU28 member states, the thrust of energy transition goals with regulatory and legal requirements for "unbundling" and the precedence of renewable energy-source electricity above fossil energy-source electricity, uncertainty regarding the FITs (feed-in tariffs), declining or stagnant electric power consumption, and large infrastructure spending needs (for example on grid systems) have in overall terms led to the utility sector being destabilized. The knock-on to the financial standing of utility companies has been significant. In 2008, Europe's top-20 market traded electric power companies bundled into European utility indices, such as Stoxx Europe 600 Utilities, the Bloomberg European Utilities Index and the MSCI European Utilities Index, had a combined market value of more than 1000 billion euros. In late 2013, the combined "market cap" of these electric power majors in Europe was about 500 billion euros.

Energy and the Economy

By taking a wider look at energy in the economy, and the factors deciding or favoring certain energy-economic processes, structures and infrastructures, rather than others, we can identify the probable or potentially most-efficient and most likely outcomes from any shock to a pre-existing system and structure.

Put another way, we can also identify the probable or likely stress points and challenges for energy transition, highlighted by stress on "historical or legacy power producers".

One particularly important lesson from energy economics for nuclear power, and other electricity supply-side sources of energy, is that intensifying the role of electricity in the energy economy has practical limits.[1] Global experience to date suggests a limit of around 20 percent to 25 percent for electricity as a part of final energy demand. Attempts at further raising electricity in final energy soon result in major difficulties, including capital costs of utility power production, transport and distribution systems.

To be sure, vested interested in the power sector, especially the nuclear lobby argue that "electrification of the economy" can be pursued to very high but usually unspecified levels. Achieving these, we can note, would for example make it obligatory to electrify the road transport sector, incurring extreme high infrastructure spending needs.

Europe's utility sector "financial meltdown" since about 2008-2010 is in many ways unprecedented. Only the post-2008 banking crisis is comparable by market cap losses, potential needs for state bailouts, and high risks for the European economy going forward. Consequently, Europe's energy transition plan is easily and often singled out as a key troublemaking factor, due in part to the rapid-rising capital intensity of European power systems, but it is very easy to argue the opposing hypothesis, that Europe's utility sector, like Japan's, has so intensely financialized itself that it is unable to react to changing energy policy and resource signals.

The nuclear lobby in Europe can hope that nuclear power may "re-enter by the back door" when, rather than if, the energy transition targets voted in December 2008 are diluted or abandoned. When (rather than if) EU energy transition goals are abandoned, or at least diluted, supporters of restarting nuclear power in Japan can claim additional credibility for their proposals.

This dilution if not abandonment of EU28 energy transition goals, of which many are "outline, non-binding or high ground", is in fact a likely outcome for many reasons, but I contend these reasons are not primarily related to the "20-20-20" goals set in late 2008 for the year 2020. In brief, these stipulate a 20 percent reduction in European CO_2 emissions and a 20 percent reduction in energy per unit GDP, relative to 1990 levels, and a 20 percent renewable energy contribution in the European energy mix by 2020.

Much more important in my view, the already-emerging dilution and stretching of EU energy transition goals is due to dysfunctionality in the economic and financial system, as well as what we can call "legacy issues". Further "upstream" in European energy, all types of energy including electricity are on a very, very slow growth path, while some types continue to decline – notably natural gas demand. Sales revenues for energy producers can therefore only rationally increase through unit price rises for energy. Where this revenue growth is not available, utility power producers can be forced to radically cut investment and employment, further hindering successful energy transition. Where increases in energy prices are nodded-through by governments, this tends to further reduce demand growth, a classic "negative feedback" process. Finance-sector expectations in the energy sector are we can note "traditionally" based on and related to "the growth paradigm".

Only by understanding the dysfunctionalities of energy transition in Europe can we understand why, in Japan today, the abandonment of nuclear power is able to be treated as controversial, difficult, and even "dangerous for national economic wellbeing". More precisely and already proven, the abandonment of nuclear power in Japan will be dangerous for the financial wellbeing of many utility companies.

At the highest level, this is finally due to defects in the political process – which are summarized as "dispossession" by Majia Nadesan.

Nuclear Debt, Energy Prices and Policies

Japan, like Europe and the US, faces severe national banking, financial and monetary stress following the 2008 global financial and economic crisis. Certainly in Europe, the financial crisis of the utility sector can be easily seen as a subset of the national and continental financial-economic crises. Within the utility power sector, nuclear energy exhibits the highest levels of financial stress, measured firstly by nuclear debt. Like the European banks, the power sector remains vulnerable to another financial crisis. As in the banking sector, capital buffers are too small, state-level creditor bail-ins are usually too small and come too late, and emergency resolution mechanisms as well as massive power price rises are the default result.

As of early 2014 and for Europe, banking "safety nets" are still inadequate to insulate governments from losses that could arise from a system-wide failure, and sector specialists estimate that another banking crisis today would cost European sovereigns between 2 percent and 10 percent of GDP. The final bill would depend on many factors, the size of the initial losses being only one. At a smaller level of course, the same analysis and conclusions directly apply regarding the financial burden of nuclear power.

We can note that state bailouts to the power sector have 'traditionally' and mainly concerned nuclear power. In Europe, the UK is currently and officially pursuing both an extreme high cost "new build nuclear" program, and at the same time, the construction of the world's largest and most expensive offshore windfarms. With no surprise at all UK power prices, which are already high even by European standards in early 2014, can only increase much, much faster than general price inflation as measured by national CPI.

Extreme high power prices are guaranteed by the UK Government to the builders of new nuclear plants in the UK, and to be sure to the developers and operators of offshore windfarms. Called the "energy strike price of contracts for difference", for nuclear electricity, this guarantees extreme-high electricity prices and therefore revenues for nuclear power plant builders and operators - which include no British builders due to the rampant deindustrialization of the UK. Similarly, all major mill equipment and offshore technology for the UK's proposed offshore windfarms will be foreign-built, when or if the already "troubled and controversial" offshore windfarm program is executed.

The first two "tranches" of new-build NPPs (nuclear power plants) in the UK will be built by France's EDF. We can note that the debt of nuclear power producers in the developed countries (both state and semi-private entities) is always high or extreme-high. Among the highest in Europe we find France's EDF, sourcing 84 percent of its power production from nuclear energy, covering roughly 75 percent of French electricity supply (and about 24 percent of final energy demand) in 2013. EDF's ballooning corporate debt is now approaching 40 billion euros and can only grow, because at present it excludes future nuclear plant decommissioning and "safestor" costs for the 63 large civil NPPs owned by EDF, the only owner-operator (by government decision) of NPPs in France.

The French general accounting office (Cour des Comptes) in a 2012 report suggested these reactors would incur decommissioning costs from 2025 of "about 3.8 billion euros, each". Power price hikes to claw the cash to start decommissioning funding are now widespread across 'Nuclear Europe', in those EU28 countries with sizeable numbers of NPPs. This round of power price hikes is unsurprisingly led by the most nuclear-committed country in Europe, France.

With total support from the "socialist" French government of Pres. Hollande, which firmly limits state sector pay rises to 1 percent a year, EDF from 2013 is hiking power prices 10 percent each year for the next 3 years. Massive power price hikes in France are above all designed to enable the constitution of a fund for NPP decommissioning. So-called "fuel poverty" can only spiral in France.

In Nuclear Europe, we find that state aid to the mostly privatized, partly "unbundled" power sector features near-automatic state support to raising electric power prices and power transport and distribution charges to final users. We also find this trend is especially strong in those countries sourcing the largest amount of their electricity from nuclear energy.

The conclusion that nuclear power is not only dangerous and weapons-proliferating, but is also very expensive, is very easy to make. The exact opposite is the basic theme of nuclear lobby propaganda.

Pricing Electricity Out of the Mix

As noted above and worldwide, electricity is a small, sometimes extreme-minority "player" in the energy mix. Even among advanced-industrial states (also called "postindustrial" due to outsourcing and deindustrialization) such as the European states and Japan, electricity rarely covers more than 20 percent of total national energy. For the EU28, using Eurostat data, the average role of electricity in final European energy demand through 2007-2012 was exactly 20.00 percent. Only in exceptional cases such as France, do we find electricity covering more of final energy demand (about 24 percent for France).

In the majority of developing and emerging economies, electricity's role in total national primary energy is typically well below 10 percent, of which, in these developing and emerging economies, nuclear power typically supplies almost no (0 percent-1.5 percent) part of national electricity.

Extreme price rises in the electrical minority sector of the energy economy can therefore only operate for a certain time, before further depressing electricity's role in the energy economy and incurring such clear or manifest system-wide economic losses that a policy of high electricity prices will be terminated. When or if the "full life cycle costs" of nuclear power are counted, including all upstream and downstream costs, uranium mining including mine development and infrastructures, NPP costs through the full lifetime of plants, related facilities, fuel reprocessing and waste treatment, plant decommissioning, and so on, the obligatory need for high electricity prices when nuclear power is chosen, becomes starkly clear. To be sure this fact is actively disputed by the nuclear lobby.

The problem is that in Europe, unlike Japan (although the difference is shrinking), electricity prices are already among the highest in the world. One striking example using Eurostat data for present (late 2013) average household supply prices in Europe's leading country for energy transition, Germany, is about 25.3 euro cents per kilowatt-hour (kWh) as of late 2013. This prices German household power at around 404 euros or 544 US dollars per barrel equivalent of energy at early 2014 USD/EUR exchange rates.

As we know, oil prices at 100 US dollars per barrel are communicated by the media (if not by the major brokers and bankers operating oil and energy markets, such as Goldman Sachs, JP Morgan, Barclays, Soc Gen and Deutsche Bank) as being "very high". Europe's industrialists repeatedly warn what extreme-high electricity prices mean for the "competitiveness" of European industry – a slogan now also heavily utilized by European politicians. With little surprise, the same arguments are used in Japan by the local nuclear lobby.

Electricity prices in Europe will nevertheless rise further and must do so in the nuclear-committed countries to finance NPP dismantling, decommissioning and safestor, as well as NPP new build. One sure and certain impact will be a further decline, or very slow growth of the role of electricity in final energy demand and consumption. In turn, this will further reduce the claimed "urgent need" to build NPPs due to the unlikelihood of "fast growing power demand".

Regulatory Shock and Energy Transition

Apart from the dream of an all-out shock transition away from fossil fuels, "to save the planet from global warming", which is enshrined in European national and Union energy legislation and policy since December 2008, Europe is the focus of an ongoing, 30-year ideological struggle by governments and the Commission to deregulate energy markets, privatize state-owned energy companies, and "unbundle" or break-up previous combined electricity generation and transmission-distribution entities.

The claimed aim of this ideological quest was (or is) the neoliberal tweet of increased competition, higher efficiency and lower energy prices for consumers.

The exact opposite is the real world result – powerfully aided by the nuclear legacy – which in Europe is a key negative factor or additional cost and charge, for energy transition.

What consumers and users have got in Europe was and is higher prices. Taxpayers will also be "tapped" to bail out destabilized and weakened utility power companies on a probably long-term and ongoing basis, when or if these financialized companies are exposed to serious financial stress. In addition, power brownouts and blackouts on an increasingly frequent basis are threatened, due to the technical, industrial, financial and economic crises converging in the power sector, notably the financially-impossible goal of developing both smart grids, for power distribution, and super grids, for power transmission. This is a "perfect storm".

Policy goals for energy transition in Europe are similar to the supposed "consensus policy goals" for European banking and financial services - the neoliberal slogan of deregulation and privatization rules. Since 2008, as in the USA, the deregulated banking and finance sector crisis in Europe is of epic proportions, and is also a continuing major threat for economic stability in Japan. Adding the slogan of "unbundling" to the slogan pack for Europe's power (and gas) sector energy policy was easy, but executing this "unbundling" has massively increased the financial and technological or industrial vulnerability of existing "legacy" power utility companies.

Renewable energy in Europe has been in some cases (like German and Spain) very rapidly ramped up to take a growing share of the market and in the above-cited countries accounts for a huge slice of total "nameplate" national power generating capacity – depending on time of day and date in the year, as much as 55 percent, as of late 2013. This at least proves that in "pure technological terms" it is possible to quickly ramp up renewable-based electricity supply and back-out fossil-based electricity supply on an "instantaneous basis" but not year-round basis, where the role of renewable power necessarily falls sharply and is both difficult and expensive to raise above about 25 percent of power generation year-round.

Coal has been almost exterminated for power generating except as an ironically growing stopgap in Europe, and EU gas prices remain at about four times US domestic gas prices, helped by anti-fracking legislation in Europe – while imported shale gas produced by fracking, in the form of LNG is seen as perfectly normal in Europe.

Electricity pricing, after firstly becoming a political football, has become a plaything of rapacious energy bankers, brokers and traders who drive peak prices ever higher, and baseload prices ever lower. To this anarchic hammer blow for energy transition, Europe's emissions trading scheme, the ETS, possibly in its death throes, has added its own destabilizing influence to power production and transmission economics in Europe. These (and other) dysfunctional elements in Europe's energy transition are surely clear risks for Japan's energy transition, needing careful analysis and action designed to avoid reproducing the "European energy transition crisis".

Regulatory action has also both hindered and intensified the range of new, developing and potential technologies for the power sector – all of them disruptive – which have crowded onstage, notably the concept of wide-area smart grids. Europe's aging and undersized real grids – the continent's "legacy" power transport systems – suffer from long-term and basic under-investment and are not adapted to increased amounts of intermittent-supply power from renewable sources. Serial brownouts and blackouts are a sure and certain threat. Regulatory action in this domain, such as forcing utilities to maintain large but little-used "spare" generating capacities, is obviously disruptive for utility company financial stability. Taking only power transport and distribution investment needs, these may total as much as 750 billion euros by 2020, according to sector-specialist entities such as Eurelec and agencies such as ENTSO-E, but funding of this mega-investment is currently impossible, either by the power sector or by governments.

One highly ironic but comprehensible effect of the brownout threat, and high power prices, is the increasing move by large industrial entities towards "self-generation", that is cutting themselves off and away from national power grids. This "forced-autonomy model" is most certainly a source of major opportunity for city authorities, local associations and small enterprises, but again may be hindered by regulatory conditions in most European countries, which limit the ability of midsized (neither small nor large-sized) producers of intermittent power to distribute and sell their output.

Faced with extreme investment needs, European power transporters and distributors are obliged to radically increase their charges. As of October 2013, Germany's power grid operators boosted the surcharge household consumers pay for renewable energy by 18 percent to a record 6.24 euro cents per kWh for 2014 from 5.28 euro cents in 2013, in a 400 percent or five-fold increase of power transport charges since 2009. In Japan, the threat of similar cost spirals is highly likely, unless alternate and different models and paradigms are applied for energy transition.

Policy Disaster – Financial Disaster

The "unbundling" and privatization theme, or slogan in policy and ideology terms is part of the globalization paradigm. This is simple to understand and apply in most industries, but these do not include electric power. This paradigm, originally coined by Ricardo in the early 19th century advocated the exploitation of different national and regional natural resource endowments, and labor costs, as well as then rapidly-developing international marine transport, is now used to defend unbundling and privatizing national electric power systems. To be sure, physical manufacturing can be delocalized or offshored to "low labor cost export platforms", to exploit cheaper labor in the emerging economies, but this paradigm does not apply to electric power on a prima facie basis.

For electricity in Europe, despite the science-fiction idea of producing non-fossil hydro and geothermal power in Iceland, solar power in the Sahara desert or (less fantasist) hydropower in Norway and transporting it to Europe, globalization is a non-starter. The European transition plan therefore attempts the next best thing of "continent-wide" electricity transport and trading, as a policy goal for "the longer term". Because this is a purely theoretical construct, it would imply continent-wide HV transmission able to transport as much as 90 to 100 GW at any one time. Merely the construction time needed for this Herculean effort would be measured at 25 to 40 years, when or if the funding was found. Therefore, in the short term, unbundling only serves the elite goal of breaking-up integrated power production and distribution entities, usually national-owned, often dating from 1945, on the theoretical basis that "increased competition" would or should reduce prices and improve efficiency.

Privatization of the electric power sector in Europe has had very variable and usually low degrees of success to date. Even when the privatized entities become the 'too big to fail' darlings of the political elite, especially the part-privatized nuclear power sector, and the state holds 'golden shares' in these part-privatized but fragile corporate structures, the companies are often very poor stock market performers and their stock prices are easily pushed lower by the brokers whenever there is a retreat in equity markets.

In a repeat of the financial market crash of the 2008-2009 type, the European utilities will very surely be some of the most spectacular victims, possibly losing a huge amount of their remaining market capitalization, while other sectors could scrape by with a 33 percent loss. As already noted, this "disaster scenario" for European utilities will also be a disaster for energy transition.

Before the crisis of 2008, and for European nuclear power before the 2011 Fukushima disaster, the continent's utility companies were stable and confident, and invested accordingly – but as noted this was on the basis of highly over-optimistic power demand and economic growth forecasts. European utilities also conformed and complied with Europe's climate-energy policies and legislation, notably by investing in gas-fired power plants, but often not in the renewables. Today for example, Germany's four-largest power companies (E.On, RWE, Vattenfall, EnBW) only account for about 6 percent of Germany's total power supplies from renewable sources, but the total peak generating capacity from renewables in the two leading countries for renewable power capacity, Germany and Spain, has mushroomed to a peak in the region of as much as 55 percent of national "nameplate total" generating capacity, depending on time of day and period of the year.

Industrial development of the new renewables has been phenomenal in Europe, but this was a policy-driven "no feedback" process, resulting in huge overcapacity, company bankruptcies, forced restructuring, job losses, investor retreat and shareholder losses. On the ground however, on windy and sunny days of low demand – the summer Sunday paradigm – the result in Germany and Spain is rampant overcapacity of power capacity, resulting in the shedding of unusable output. The argument that transmission and distribution capacities can be rapidly increased, and utility-scale power storage can be developed at some stage in the future, is no answer to actual and current dysfunctionalities.

Investors at first trooped to place funds in renewable power production in Europe, lured by the very generous FITs and prime rank allocated to renewable producers, first in line for supply to distributors. Until about 2010 this was a win-win prospect, resulting in literally massive growth of European renewable energy equipment production – an industrial and technological triumph. But the blowback from the ETS shambles in Europe, extreme and continually rising power prices for consumers hitting demand, near-saturation of prime onshore windpower sites in several countries, the huge fleets of mothballed gas power plants weighing heavily on utility company balance sheets, corporate debt or "leverage", and the post-2008 banking and finance crisis combined to implode investor confidence and interest in the power sector.

Power traders added more stress to the balance sheet through bidding down baseload power prices – made rational by rampant overcapacity – often slashing baseload prices by 50 percent in 5 years, for example in Germany where baseload power in late 2013 was priced, some days, at 38 euros per 1000 kWh but typically cost more than 80 euros in 2008. For German household consumers, of course, these numbers are derisory because they pay at least 6 times, and sometimes 8 times that price. Their likelihood of consuming more electricity is therefore zero, like the ability of power companies to attract huge investor inflows to pay for often extremely high-cost investments, notably in grid development, urban smart grids, and combined heat-and-power projects.

Energy Policy Vacuum

Across Europe, five short years ago, the call for "energy transition" or transformation was deafening. Since then, the onrush of renewable power capacity has done a lot more than simply put pressure on utility company margins - it has destabilized and transformed the established business model for utilities. Whatever the power production and TD (transport-distribution) sector was, previously, it has now radically changed and is both destabilized and dysfunctional.

As already mentioned, investors hate uncertainty – and they despise utility companies with a dreaded combination of declining or negative margins, high debt, huge investment needs, and a very uncertain future. The utilities themselves have massively "migrated" to financial engineering risk-cover, putting their heads into several silky nooses. In particular, power hedging is now widely used – typically selling up to two-thirds of their power output one to three years forward. Today therefore, they can receive 2010 power prices, insulating them from a lot of the general decline in prices since 2010, but when these contracts expire in the next 12-15 months, the financial damage may be large or extreme.

The financial retreat from the power sector was impossible to predict, even in 2009, but has now snowballed. The slogan and rallying call of the financial industry – unbundling and privatization, or "deregulation" – was massively applied to European electricity by Commission mandarins and national governments, but along with the renewables, it swept away the previous ordered and predictable system, for utilities, of producing power according to the marginal cost of generation.

In Europe today, and possibly a model that Japan should not adopt or should heavily modify, renewables rank highest and have grid priority, by law, meaning the grid must take their electricity first. This can be defended by completely classic economics: Because the marginal cost of wind and solar power is zero, grids would normally take their power first, anyway. The fly in the ointment, or elephant in the closet, is that wind and solar power are impossible to treat as baseload power – meaning that the whole power generating structure and system becomes intermittent.

The European power sector is now "transformed" but in no way to the taste of investors, banks, pension funds and brokers – or consumers. The political impact of this has been steadily mounting and will continue to do so. The previous "hands off" attitude to energy policy adopted by the majority of EU28 governments will certainly change – but as of present it is hard to identify the weights of the different elements in the complex of factors weighing on EU energy policy. One radical change, which is possible, is renationalization of the power production and TD sector, very possibly not called that due to sensitivity on this "Soviet-minded" political action. To the extent that "neoliberal-minded" tinkering with the power sector has already produced major, even systemic dysfunctionality and extreme high costs for consumers and voters, we could argue that renationalization would have problems in being a greater evil.

Most important, and certainly comparable with Japan post-Fukushima, Europe has entered a policy vacuum concerning electricity, as well as the political meanings of and measures for energy transition. One danger is this vacuum enables political deciders to not decide, while hoping the problem "goes away", that is traditional muddling through or "kicking the can down the road". In turn this will however not be possible due to the already-massive changes that have occurred (both by design and by accident) to European power in at most five or six years.

Saving the Dinosaurs

The utilities balked at the culture shock of "getting serious about renewable energy", and in many EU countries have gone into a form of denial. The green and ecology parties and NGOs accuse them of behaving like ostriches – or dinosaurs. They say the power sector uses dinosaur fuels from the Jurassic and earlier, and made itself extinct like the dinosaurs.

Corporate strategies have been put through the meat mincer, often several times over. Some utilities are attempting to shift away from producing power by moving downstream, further increase their energy trading, supply consulting services on customer energy needs, and possibly invest in energy storage and smart grids (despite the costs), while also cutting back on their total generating capacity, sometimes quite radically. Several are moving out of Europe, to produce electricity in more favorable and more predictable climes. Germany's fourth biggest utility, EnBW, has even come out with a late-2013 forecast that its earnings from power generating will fall by 80 percent in the eight years 2012-20, offset by hoped-for higher earnings from energy services and from a belated shift to renewables.

Above all however, the staggering loss of market value suffered by the 20-largest power sector companies in Europe is a warning signal. Their losses have a somber meaning for their directly-employed workers still numbering more than 1.25 million in Europe, with a very high multiplier for dependent jobs, as well as for pension funds and other investors. The hoped-for transfer of employees from old model power sector companies to the new green model is very far from sure, for reasons as basic as the huge and rapid loss of capitalization suffered by the power sector, and the policy-driven boom-and-bust cycle that has afflicted the renewable energy industry of Europe. While the utilities can be blamed for their slow reaction to inevitable change, governments (and in Europe the Commission) can also be squarely blamed for a policy rout due to completely incompatible elements being thrown together in a "dog's dinner" of expedients and stopgaps.

It is for example certain that power sector companies will have to pay their workers and shareholders less, in a context where governments will get less tax revenues from the downsized utilities, and be forced to bail them out like Europe's "bad banks", while having to maintain feed-in tariff and other subsidies to renewable power. The poor earnings, and often astronomic corporate debt of the utilities – especially for their nuclear "assets" (we mean liabilities) – results in their bond prices being hammered as their borrowing costs soar. Even the largest power utility and TD companies in Europe are now forced to pay rates of 10 percent a year on their borrowings, while the ECB pushes prime bond rates in other sectors to almost zero percent. The utility sector is "risk on".

The abrupt decline in the power sector's fortunes can only call into question its "new role" as clearly set out in massive numbers of national government and Commission policy papers. The utilities and the TD companies were going to be the "suppliers of last resort", guaranteeing the lights never went out, and were going to be the lead investors and pioneers of the all-green transformation. Today however, it is very rash, even plain foolish to imagine this is possible and will happen.

What Japan Can Learn from the European Mess

Only for the moment is Europe's "green magic" working, that is, green power is being ramped up and the lights are still on, even if the consumers and users are grumbling about how expensive electricity is getting. The claim by green political parties that this itself is a proof that other politicians and the public are over-worried about the risks of energy transition is very easy to challenge.

When or if national, or even semi-continental power grid and supply brownouts and blackouts regularly occur, and when – not if – electricity prices go on rising, we can be certain that open and naked political handwringing and soul-searching will occur, even in the most government-friendly media. The power sector may well be a handy whipping boy for politicians and the media – they either overinvested, or invested badly, and were not sufficiently "pro-active" in Europe's green power adventure, and so on, but this boils down to a political problem which has to be faced.

Japan can for example treat the removal and substitution of its nuclear generating capacity as a separate and definite goal, and take all needed steps to prevent the actual results of European energy transition – notably that power sector companies are shedding capacity, and at best making very low levels of new industrial investment, because their corporate finances are so weak and the outlook is so uncertain. Japan must strive to prevent the emergence of a context where the utilities have every corporate financial reason to not invest, and to divest. Japan must avoid the de facto result in Europe of the utilities making a so-called "strategic retreat from their core business" of producing and distributing power which is not in any way – except the most disruptive – being offset and succeeded by smaller new entrants.

The needs cannot be met by "conventional" means. Alternatives need to be put on the table and they will certainly include political and policy alternatives.

Also very similar to the European situation, where governments are saddled with astronomic sovereign debts and very slow economic growth, Japan may decide the state should get a return from the public money used to bail out the power sector – unlike the public funds repeatedly used to bail out the "bad banks".

In other words, as in Europe, Japan has an option of renationalizing the power sector with the aim of making it a source of government revenue – not additional debt, earning some money for the state's shareholders – the taxpayers and voters.

This will of course generate howls from "free marketeer liberals", whose pantomime liberalization of the power sector is one of the major causes of its dysfunctionality in Europe. Swift and decisive government action in Japan, as in Europe, could defuse and expedite the layer cake of crises that has so rapidly and dangerously built up in the anarchically privatized-deregulated power sector. Creating new national energy companies with public investor subscription may be a popular success given the existing and growing problems for maintaining public support to energy transition.

Other options include increased and highly rational energy decentralization, for example using the model of German's stadtwerke or municipal energy, water and heat providers enjoying a high level of public confidence. Renewable energy is above all decentralized, therefore its production, supply, pricing and management should also be decentralized. Specifically for the power sector, dedicated new sovereign wealth funds might be created, or existing ones oriented to this sector. In-company supply and industrial self-generation have in any case moved ahead rapidly in Europe – and will almost certainly grow in Japan due to high or rising power prices and declining reliability of supply. This process should be formalized and structured through the right combination of public-private policies, rather than being allowed, by "laisser faire", to cause further power sector disruption and larger and further earnings losses for power utilities and TD companies.

For the moment, in part due to Europe's banking, debt and economic crises and also because of the fantastic pace of green energy transition, the power sector is being sacrificed. Estimates made by the German environment ministry in late 2013 suggested that since 2007, total European spending on "green energy transition" may have attained 900 to 1000 billion euros, but the ground-level results often include stark failures and a grotesque waste of human, technical and economic resources.

Japan therefore needs to look carefully at Europe's confused energy transition policy, programs, measures and spending before setting its own national transition policy and programs. Either within this, or apart and separate, Japan can also set the goal of eliminating its nuclear power capacity with confidence

Note

1. The figures for 27 EU countries are at: Sources/Entities: U.S. Census Bureau, Population Division - USA; Statistics Bureau, Ministry of Internal Affairs and Communications - Japan, Eurostat / IEA / UNECE / National Entities, Eurostat / UN / NSI, PORDATA at http://www.pordata.pt/en/Europe/ Final+energy+and+electrical+energy+consumption+per+capita-1732-102486

The Nuclear Energy Paradigm Collides with Earth Changes and Technospheric Breakdown

The Fukushima Five

The title of this chapter does not really represent its most important message. Undoubtedly, Earth changes are upon us in the form of increased seismic activity and volcanism. *Global Climate Change* is intensifying on every continent. And technospheric breakdown is accelerating by the year.

However, that the worldwide *Nuclear Energy Industry* (NEI), and the governments which regulate it, proceed down the same road despite these changing realities points to one thing in particular. As follows:

The voice of reason has fled humankind, especially within the scientific establishment which informs the Nuclear Energy Paradigm (NEP).

Likewise, common sense has become quite rare throughout the governments and corporations of the world where nuclear power generation has been promoted.

Because of this unfortunate state of affairs, critical decisions about the outworking of the *Nuclear Energy Paradigm* have been made, which have put whole populations around the world in great danger. In the same way, nations such as Japan have risked their future for the sake of clinging to an energy source which no longer makes any sense, nor ever did.

Why would the people of Japan put their four major islands in such jeopardy? "What jeopardy?" the Japanese politicians and officials at TEPCO ask.

Just prior to the M9.0 earthquake and subsequent tsunami on March 11, 2011, Japan had 55 nuclear reactors in operation throughout the country, with an additional 12 approved and in various stages of development. Despite the fact that Japan is located in one of the most seismically active regions in the entire world, TEPCO and the Japanese government have relentlessly proceeded in the direction of nuclear power generation to meet the nation's energy needs.

Our journey into the minds of those critical decision-makers begins here. Particularly in light of the fact that Japan was already the site of two nuclear disasters during World War II, one would think that the national and corporate leadership had been forewarned. And yet they proceeded to site virtually every nuclear power plant on the coastlines of the four main islands, exposed to all sorts of earthquakes and tsunamis. (See Graphic [1] in Graphic Links below)

First let's establish that there is a huge difference between the risks and dangers associated with nuclear energy and those that simply don't exist with hydrocarbon, hydroelectric, coal, solar and wind power generation models. When those power plants get hit by an earthquake, or overwhelmed by a tsunami, they simply shut down. There may be serious localized damage, but it's usually very easy to contain and straightforward to fix.

Most importantly, such inevitable Earth change events will not precipitate a global catastrophe, as was the case with the Fukushima Nuclear Disaster. Whether by way of typhoon or tornado, earthquake or volcano, tsunami or flood Japan's nuclear power plants, especially in view of their locations, will always remain extremely vulnerable. Hence, it can be stated with complete certainty that the responsible decision-makers committed a huge blunder by taking Japan in the direction of nuclear energy in the first place.

How has Common Sense Become Rare and the Voice of Reason Muted?

This entire chapter could easily end right here ... end of discussion for any rational person who simply applies human reason to the predicament of Japan. How so? When a good answer cannot be provided to the following questions, what else is there to discuss about the viability of nuclear power generation on the most earthquake-prone island nation in the world?

1) What was the contingency plan at Fukushima, and for all the other nuke plants throughout Japan, should a tsunami overwhelm the plant buildings and completely cut off the electricity supply? When a long term disruption occurs, how will the reactors be cooled down?

2) How is a nuclear meltdown to be prevented in the wake of radioactive explosions and exceedingly high levels of radiation measured throughout the entire local area? How is the safety guaranteed of all the emergency personnel tasked with making the repairs necessary to prevent a meltdown or worse?

3) What was the TEPCO contingency plan in Fukushima should the power supply be cut off from the cooling pools where the spent fuel rods were being stored? What are the standard operating procedures in place throughout the rest of the nation's nuke power plants?

If there is not a reasonable and actionable implementation plan for any of these predicaments, should they occur, then the plants simply shouldn't be there. And, yet, there they sit -- without any discussion of these and many other catastrophic contingencies -- while the politicians in Tokyo talk about the coming Summer Olympics in 2020. It's important to bear in mind that Tokyo is a mere 180 miles from Fukushima, and that some of the Olympics venues are even closer!

When the Faculty of Human Reason Fled Humankind

Not only does no one really know the extent of the environmental damage left in the wake of the Fukushima nuclear disaster, no one knows how it might further develop in the future. Nevertheless, the government persisted in their pursuit of the Olympics to be played out right in the backyard of perhaps the greatest nuclear catastrophe in history. The decision process in both cases cannot be driven by simple logic or rationality.

If ever there was an event which exposed a clandestine plot going on behind the scenes for several decades, Fukushima is it. Certainly those who have directed Japan's destiny failed to take into consideration the obvious: That Japan occupies the most earthquake-ridden piece of real estate on the planet, and that *Global Climate Change* does not discriminate; it can show up on anyone's doorstep uninvited and when least expected.

What the whole world has witnessed since that fateful day in 2011 is one earthquake after another, together with a number of severe storms buttressing the four main islands. In each instance, all of Japan's neighbors -- friends and foes alike -- watch with baited breath for the many and various outcomes. Particularly in light of how precarious the situation at Fukushima continues to be, is the entire community of nations wondering if and when this intractable situation will be resolved? Clearly, we shouldn't even be here, and wouldn't be if those responsible had applied even a modicum of common sense during the incipient stages of development of Japan's nuclear energy policy.

Fukushima Feels the Brunt of the *Perfect Storm*

Fukushima is a perfect example of exactly what can go wrong, particularly when the *Perfect Storm* blows throughout the area. Just as Superstorm Sandy revealed just how old and antiquated much of the infrastructure is throughout Lower Manhattan, the March 11th earthquake and tsunami of 2011 allowed the many manmade problems and inherent weaknesses of the Fukushima Daiichi Power Station to surface over the course of the following weeks. Each day of the unfolding nuclear disaster seemed to bring a new development, none of them being positive. Some of these were expected; some came out of left field.

At the end of the day it was obvious that no power plant could withstand an M9.0 earthquake and 45-foot high tsunami (See Graphic [2]). However, even much lesser events could take greater tolls on such a vulnerable location as Fukushima ... or Tokai (Tokai Village)... or Onagawa, the sites of other nuclear power plants. The moral of this story, and it is worth repeating until Japan -- the whole nation -- gets the message, is that they had NO business building these nuclear power plants along one of the most seismically active coastlines on the planet.

An article titled "Magnitude 6.8 earthquake off Japan coast triggers tsunami" (Associated Press) appeared in the media as this chapter was being written. Truly another telling sign that this whole predicament can be quickly duplicated anywhere along the Japanese coastline where nuclear power plants are located.

In addition to the constant threat of earthquakes and tsunamis, there is also the growing number of typhoons (Kageyama, Y.) making landfall on the Japanese coast. The increasing intensity of these storms is putting them in the same category as Superstorm Sandy. Therefore, they will undoubtedly serve to reveal more problems with the nuclear plants that dot the Japanese coastline should they receive direct hits in the future.

Global Climate Change Reveals Fatal Flaws of the Nuclear Energy Paradigm

While Fukushima is the poster child for everything that can go wrong -- very wrong -- with nuclear power generation, there have been several instances since March of 2011 which have highlighted the inherent weaknesses found in the global *Nuclear Energy Paradigm*.

Photos often tell the story much better than all the words in the world. The media carried startling photographs of Nebraska's Fort Calhoun Nuclear Power Station during the floods of June, 2011. (See Graphics [3 and 4]) The plant was completely surrounded by floodwaters and under threat of losing its power supply. Even though it had been in cold shutdown since April, it was still vulnerable to problems not too dissimilar to Fukushima Daiichi plant.

As most of us know, the mainstream media rarely reflect the true gravity of either natural or manmade disasters. As an example, for days immediately following the M9.0 earthquake the media repeatedly reported a death toll of between 100 and 200 people after the tsunami struck the northeast coastline of Honshu Island on March 11, 2011. Certainly the authorities knew very quickly the devastation that the 45-foot high tsunami had inflicted. That whole towns and coastal villages were swept away became common knowledge after the first flybys. So was the fact that tens of thousands of Japanese citizens had been killed by this tragic event. Some of the most authoritative accounts have reported that upwards of 30,000 people passed away from the tsunami and aftereffects which lingered for months due to an exceptionally unresponsive government in Tokyo.

What's the point? The governments (and the press) are both conditioned to grossly understate the seriousness of any matter which impacts so many citizens in any given area. Therefore, when the big media outlets do trumpet a genuine concern about the integrity of the nuclear power generation model, as they did in the wake of the Nebraska flooding, it can be safely assumed that some critical issues have not been thought out. Hence, we saw headlines like the following:

Flooding Brings Worry About Nebraska Nuclear Plants (Sulzberger and Wald)

Fort Calhoun Nuclear Plant: Flood Seeps Into Turbine Building At Nebraska Nuke Station (Funk and Lampe)

Flooded Nebraska nuclear plant floods, raises broader disaster fears (Hargreaves, S.)

The US Is Having A Catastrophic Nuclear Emergency In Nebraska And The Obama Administration Is Covering It Up (Blodget, H.)

Fort Calhoun Nuke Plant Flood Wall Collapse the Result of Accidental Puncture (The Doc)

There is one central and critical point which the preceding photographs graphically portray. Because of the ongoing intensification of *Global Climate Change*, this scenario can repeat itself virtually anywhere around the globe. Wherever there is a nuclear power plant, it can be overwhelmed by floodwaters if the rains and subsequent flooding are severe enough. When the plants are located on the coastline, as many are, to provide easy, inexpensive access to ample water supply, the threats greatly increase due to hurricanes and typhoons.

The real question remains, therefore, as to what type of contingency plans the various nations of the world have in place to address the worst case scenarios. The *Perfect Storm* that occurred at Fukushima can easily be duplicated in the form of a catastrophic weather event that temporally paralyzes a nuclear power plant, especially one that is in the way of raging floodwaters or a category 5 hurricane or typhoon.

It's almost as though Mother Nature endangered two, not just one nuclear power plant, in 2011 as a way of showing the whole world that every plant is just a superstorm, earthquake, flood, or tsunami away from a full-blown nuclear disaster. In view of the government and corporate response during the Fort Calhoun flood, it's clear that circumstances can, and do, arise which present extremely challenging predicaments. Sometimes they can be waited out; sometimes they can't, as we have all seen with Fukushima.

Nuclear Energy Paradigm Fatally Flawed from the Start

The real problem here is that the very premises and assumptions underlying the NEP are fundamentally defective. It's especially important to point out that the very scientific and technological foundation of this paradigm is woefully lacking. Particularly in light of the calamitous consequences that will occur when things go wrong, it is fair to say that the profound operating deficiencies and material defects of the NEP should have been significant enough to have prevented the nuclear power industry from building so many plants so quickly… everywhere on Planet Earth.

When a hydroelectric plant is incapacitated due to a storm, flood or earthquake, it only means the temporary loss of electricity generation. It is the same for a coal-fired power plant or biomass incinerator. When nuclear reactors are shut down on an emergency basis, and infrastructure is seriously threatened or damaged, what reasonably can be done? Particularly when the building and surrounding areas are permeated with escaping radioactive isotopes, even a meaningful first response can be drastically thwarted.

Of course, depending on the damage sustained by both building(s) and reactor(s), scenarios can easily develop which defy any kind of solution in both the short and long term. There are also instances, such as Fukushima, which will haunt a nation for generations to come, and in ways that are yet to be known or experienced because of the evolving complications.

At the end of the day, it will be completely understood that the state of the art (and science) which undergirds the nuclear energy industry is terribly underdeveloped. The design and engineering, architecture and materials are woefully substandard, especially when these bump up against Earth changes in the form of earthquakes or extreme weather events. Knowing that so many obvious deficiencies existed -- "all over the place" -- has guaranteed a future blighted by nuclear accidents both seen and unseen. Incidentally, it is the unseen and therefore unreported nuclear mishaps which can be the most dangerous to human life and property.

Technospheric Breakdown Comes of Age ... and with a Vengeance.

Technospheric breakdown is a concept that most are not familiar with. It is rarely discussed in the mainstream media because of what it truly means for the entire planetary civilization. As a short description, consider that much of what has been built around the globe during the past 100 years belongs to the technosphere. To a great extent, it is represented by much of what has been created since the advent of the Industrial Revolution. We are using the term very loosely within the context of nuclear energy because of how critical it is to the integrity and longevity of the existing nuclear power generation paradigm.

As an example, when Superstorm Sandy blew through the US Northeast on November of 2012, it battered that regional technosphere. In so doing, it revealed weaknesses and vulnerabilities that only a storm of that size and magnitude could do. In what ways did Superstorm Sandy reveal the effects of technospheric breakdown throughout New York City? For one, the city's well known subway system went completely down from Wall Street to the Battery. Old equipment, faulty wiring, saltwater meets metal machinery, and many other issues quickly surfaced under the assaults of the downgraded hurricane. Years of neglect and old parts combined to predictably reveal vulnerabilities with the unprecedented flooding of the underground subway system. The following excerpt comes from an insightful 2013 article that reviewed the superstorm saga as it affected Manhattan's subway.

> If a major hurricane hit New York City tomorrow, damage to the subway system would likely be worse than it was in the aftermath of Superstorm Sandy.
>
> How is that possible?

That storm did $5 billion of damage to the Metropolitan Transportation Authority (MTA) systems, much of it caused by the flooding of subway tunnels.

Power outages in the wake of the storm slowed the rate at which the MTA could pump the tunnels clear, and millions of gallons of salt water took a heavy toll on tunnels filled with metal and electrical equipment.

And while the tunnels are dry for the moment, the heavy post-Sandy workload — pumping tens of millions of gallons of water over consecutive days — degraded the pumps. (Davies, A., 2013)

Looking at just the technospheric breakdown alone, without considering the ongoing and unsettling impacts of Earth changes, is where we will start in order to establish a frame of reference.

Upon assessing the typical nuclear power plant, like the one at Fukushima, what quickly came to light were the previously identified problems with the GE reactors (Washington's Blog). They were widely known to have design flaws which made them unsound and unsafe. The longer that particular design is in operation, the greater the likelihood of a failure of some sort, and depending on the failure, potentially a major nuclear event. Bear in mind that there are other design flaws in the older nuclear reactors which either haven't yet been found, or more importantly, are not talked about because they can trigger an immediate termination of licensing in the US by the Nuclear Regulatory Commission.

Again, the context that we would like to present is the one that assesses the normal wear and tear in a nuclear power plant environment. Because the industry is so young, and under so much scrutiny by some very smart scientists who have some degree of common sense, it is now understood that the stresses that exist in the nuclear plant environment are quite extraordinary, relentless as in 24/7, and highly destructive. After all, we are talking about the process of 'controlled' nuclear fission. How can the proper containment of such a profoundly destructive process not produce some unwanted unintended consequences to people, plant and the surrounding environment?

Does the Current NEP really 'Contain' the Process of Nuclear Fission?

Taking a closer look at the process of nuclear fission, it is self-evident that such an inherently violent process was never easy to truly contain. The status quo found within nuclear power plants across the planet reveals many weaknesses, particularly the closer one gets to the actual nuclear reactors, and especially the nuclear reactor core. Engineering and design flaws notwithstanding, even if the reactors were manufactured perfectly according to the original blueprint, it has now been determined that there are fatal flaws which will reveal themselves the longer a nuclear reactor is in operation (Gunter, P.).

Scientists and engineers who have monitored the breakdown of the reactors and plants over decades are now seeing just how difficult it was to foresee the many problems that are emerging throughout the worldwide nuclear energy industry. With each passing year they observe different failures and malfunctions which speak directly to the "hostility" that nuclear power generation poses to everything and everybody found within a close proximity to the reactors. Hence, all the materials and parts, equipment and machinery, that is located in and around the nuclear reactors are always subject to partial failure or complete breakdown at a much faster rate than originally assumed. Nothing -- not even the plant personnel -- is going to last the projected life span that was insinuated during the startup of the industry.

> This is a turnaround because until recently, the life expectancy of reactors was growing. When the Nuclear Regulatory Commission began routinely authorizing reactors to run 20 years beyond their initial 40-year licenses, people in the electricity business began thinking that 60 was the new 40. But after the last few weeks, 40 is looking old again, at least in reactor years, with implications for the power plants still running, and for several new ones being built (Wald, M.)

Herein lies a crucial issue for every nuclear power plant that has been in operation for over 40 years. The vast majority of nuclear reactors were designed during the 1960s and 70s. Hence, we have an industry that is still completely relying on the state of the art of that period. Even with the leaps in scientific knowledge and technological advancement over the same timeframe, the static nuclear power generation industry decided to stick with the original 'blueprints' for new projects. These blueprints, it is now known, contain serious defects and pervasive deficiencies.

Too Many Critical Decisions at the Very Top were Driven by Economics

It ought to go without saying that when playing with fire it would be wise to first make sure that it will not turn into a forest fire ... or brushfire ... or wildfire. In the case of the NEP, this golden rule was not followed. Because the global *Nuclear Energy Industry* was essentially hijacked by corporate interests at a rudimentary stage of development, the key financial decisions trumped all other more important considerations. Because important issues were taken out of the hands of the scientists and engineers early on, the design, engineering, manufacturing and construction of nuclear power plants were often driven by cost containment and profit margins, instead of radiation containment.

This single dynamic, which has operated to varying degrees throughout the industry, essentially sealed its fate. Because the actual costs associated with truly safe, efficient and high-integrity nuclear power generation are so high, the industry would never have built even a single power plant, except perhaps a prototype in order to study all that can go wrong in such a high intensity and relatively unknown environment. Little did the plant operators know that stresses -- both macro and micro -- would take a toll in ways that could not be anticipated.

In the current analysis by those "from whom reason has not fled", it is clear that it is the subtle and unseen micro-stresses found within every nuclear reactor in the world which can cause the most serious problems. Because these unrelenting micro-stresses usually occur under the radar, they often go undetected for years or even decades. Then, one day a major failure occurs with a critical part or piece of equipment. The actual breakdown was never anticipated, but occurs as a result of a long-term stress manifesting in this way or that way.

Nuclear Power Plants Use Outdated Protocols and Inadequate Maintenance

In light of the fact that there are so many unsolved and/or unresearched problems within the NEP, the plant operators don't even know what they don't know. Stated another way, willful ignorance of some of the most significant issues due to the cost of remediation has blinded the corporate decision-makers to the real and needed solutions. It pays them not to know just how catastrophic some of the 'unexpected' outcomes can be in the face of inordinately expensive yet necessary overhauls of the plant, materials and/or reactor parts.

Studies have been conducted which have conclusively shown that the actual wear and tear observed in every nuclear plant environment has exceeded what was originally anticipated. As an example of what the industry has learned over the short history of nuclear power generation, the following excerpt delineated only some of the "Metal Aging Degradation Mechanisms" which are routinely found in a nuclear power plant.

> This is a partial listing of aging degradation mechanisms for metals, with examples of effects greater than anticipated in plant design and methods to address them.
>
> **Corrosion:** deterioration of a material resulting from reactions with its environment.
>
> Some steam generator components, piping, pressure vessel internals, and other plant areas have experienced more extensive corrosion than originally assumed during plant design. Major forms of corrosion include wastage, stress corrosion cracking, erosion/corrosion, crevice corrosion, and intergranular attack.

Methods for alleviation include inspections for signs of deterioration, control of water chemistry, or replacement with resistant materials or designs.

Fatigue: the deterioration of a material from the repeated cycles of thermal or mechanical loads or strains. Some fatigue failures in piping and other components have occurred, often resulting from larger than anticipated loads or combinations with other degradation mechanisms (e.g., corrosion).

Methods for alleviation include inspections and more accurate estimates and monitoring of the magnitude and frequency of cyclic loads.

Embrittlement: change in a material's mechanical properties such as decreased ductility and reduced tolerance to cracks resulting from thermal aging or irradiation.

Some embrittlement has been found to be more rapid than originally anticipated in plant design. Neutron irradiation of reactor pressure vessels (RPVs), for example, can lead to a more rapid loss of ductility than expected, particularly when copper and nickel are contained in RPV weld materials.

Methods for alleviation include more accurate estimates of thermal exposure and neutron fluence histories and their effects, revised operations (e.g., arranging fuel to reduce neutron flux to certain RPV regions), and component replacement or refurbishment (e.g., RPV annealing).

Fabrication defects can contribute to more rapid fatigue cracking and corrosion.

Casting and forming defects and weld-related defects embedded in a material may worsen from cyclic loadings, or such defects may become exposed by corrosion. Methods for alleviation include inspections using non-destructive examination techniques to detect embedded flaws early, and repairs.

Mechanical effects include vibration, water hammer and wear. Vibration and water hammer can result from fluid flows and result in loads greater than explicitly considered during design, contributing to fatigue failures and damage to pipes, valves, and pumps. (U.S. Congress, Office of Technology Assessment)

When put into proper perspective, this particular set of material assessments and mechanism analyses is as shocking as it is alarming. That any or all of these degradations occur wherever nuclear reactors are churning away should have given great concern to those who regulate the industry. And yet we rarely hear that the regulatory bodies focus on any of these types of inevitable decay and common disrepair found within every nuclear power plant.

The reason we don't hear about these is because once the costs were sunk into every enormously expensive power plant, the only way to recoup them was to permit them to run as long as possible, irrespective of the serious safety and inevitable breakdown issues. Of course, with the onslaught of Earth changes and acceleration of technospheric breakdown, the known processes of deterioration will only increase in number and intensify depending on how exposed each power plant is.

The biggest factor, however, is how religious nuclear power plant operators are -- in each locale -- with regard to the required maintenance and part replacement schedules. The engineers have always been aware of the obvious weak points in the system, which require regular monitoring and preventative measures. In fact, the entire industry has been performing many acts of overcompensation for years to ensure that nuclear accidents are kept at bay, as demonstrated by the extraordinary work records kept when plants shut down for unplanned yet "required" maintenance.

Nevertheless, problems arise in many instances where maintenance is not undertaken, and where parts and equipment are not replaced, because those issues have not even been officially addressed by either the industry or the regulatory bodies. The local operators can easily err on the side of neglect because of the cost pressures which have continued to erode profit margins, which were overzealously projected at the outset of the industry.

The US Nuclear Power Program is Based on an Exceedingly Bad Business Model

Bear in mind that the following quote comes from a 1986 issue of *Forbes Magazine*, and that the economic and financial status quo throughout the US nuclear power industry has only sunk lower since the mid 1980s.

> The failure of the U.S. nuclear power program ranks as the largest managerial disaster in business history. The utility industry has already invested $125 billion, with an additional $140 billion to come before the decade is out – and only the blind, or the biased, can now think that money has been well spent. – *Forbes Magazine*, February 1986

Maintenance procedures and preventative measures, plant overhauls and cold shutdown, temporary plant closures and premature decommissionings have plagued the nuclear energy industry from the very beginning. It is now experiencing a snowballing of these events, as the costs associated with them are skyrocketing. What relevance does this worsening predicament have to Earth changes and technospheric breakdown?

First, both of these major co-factors -- Earth changes and technospheric breakdown -- will only intensify for the foreseeable future. As they do, the business model that undergirds the nuclear energy industry will become more untenable. Not only is nuclear power generation the most cost ineffective of all the major energy-producing platforms, it is vulnerable to the greatest costs associated with required maintenance and repair, as well as failure prevention and remediation in the wake of a nuclear accident.

Clearly, because of unrealized revenue generation targets and cost overruns alone, the NEP business model is unsustainable. *Global Climate Change* and *Technospheric Breakdown* will only create an environment wherein costs will continue to escalate dramatically as the many aging plants worldwide get older and more decrepit. Ultimately, a breaking point will be reached whereby the current form of the *Nuclear Energy Paradigm* will no longer be a reasonable business proposition; not that it ever has been.

That Fukushima has placed such an exceptional financial burden on TEPCO validates this ubiquitous and ongoing scenario. Not only has Fukushima effectively bankrupted TEPCO, it has placed a huge cost burden on the Japanese government. It also has the potential to drag down every other business concern associated with TEPCO. The Japanese government is not immune to the extraordinary claims which may be filed in the future by the countless citizens and businesses that have legitimate grievances. These unfunded liabilities alone may end up taxing the people in ways never seen before.

The international cost ramifications have been curiously downplayed in this regard. However, given the current state of tensions between Japan and some of its neighboring countries, the current compassionate stance can easily be replaced with an understanding that Japan really screwed up, and that it ought to be held liable for damages related to all Fukushima-generated radiation damage. If damages could be proven in an international court of law, both China and North Korea might have a serious change of attitude in the not-too-distant future.

Perhaps then the politicians in Tokyo will begin to respond to this matter with prudence and foresight. The current energy policy surrounding nuclear power has been a gross failure, even when based on economics alone. Since 2011, so much has happened, and not happened, throughout the Japanese nuclear energy industry that it is a wonder any money is still allocated toward its proliferation.

Technospheric Breakdown is Accentuated by Earth Changes

The point of this part of the discussion is that as technospheric breakdown accelerates, and it is now accelerating at an ever-increasing rate because of old plants and infrastructure which are approaching the end of their lifetime, the technosphere is also exposed to the intensifying assaults that come with *Global Climate Change* and other major Earth changes. Because of this rapidly evolving worldwide predicament, the entire technosphere is now much more susceptible to both systemic breakdowns and isolated failures. Of course, a *perfect storm* of multiple isolated failures significantly increases the risks for systemic breakdown. As an example, the Fukushima nuclear disaster provides various instances in which the systemic breakdown produced emergency situations, and also of where different isolated failures worked their way into the system with tremendous consequences.

Neither TEPCO nor the Japanese government is too keen on making these types of observations. When the ongoing and ever-worsening predicament of technospheric breakdown is considered in any objective analysis -- before, during or after the event -- there can only be one conclusion. The ultimate determination that one can draw from the facts on the ground at Fukushima is that the Daiichi Nuclear Power Station shouldn't even be there. Nor should any other nuclear power plants which have been sited on any of the coastlines of Japan, so vulnerable as they are to earthquakes, tsunamis and typhoons.

In this new era of unpredictable and cataclysmic Earth changes, this obvious conclusion cannot be overstated. It is also applicable to every other nuclear power plant in the world that is located on or very near a shoreline or fault line, large river bank or volcano. Clearly, the scientists and engineers, architects and designers, manufacturers and builders who drew up or contributed to the original plans for nuclear power generation infrastructure never anticipated these eventualities. Why they did not simply defies common sense. The result is this – the horrific devastation at TEPCO's Fukushima Daiichi Nuclear Power Station (See Graphics [5], [6] and [7]).

Japan has Had Different Reasons to Flatly Reject Nuclear Energy

Particularly in light of Japanese sensibilities, as well as its poignant history, it is quite shocking for the rest of the world to watch their plight. Japan's neighbors -- both near and far -- have watched in horror as this nuclear disaster has evolved. Those nations have been concerned about the developing outcomes at Fukushima, not only for the welfare of the Japanese people, but also of their own citizens, which brings us to the point of social responsibility.

Decisions such as instituting a national energy policy that revolves around nuclear power generation can no longer be made in a vacuum. Japan, being an island nation, has always been somewhat isolated. Culturally, they are a very homogenous people and therefore tend toward groupthink. However, the Japanese government is now obligated to think outside the proverbial box. Their actions -- and inactions -- can and do have far-reaching repercussions for their neighbors, as well as the rest of the world.

When the potential consequences of Japan's actions are as grave for the Pacific Ocean community as the Fukushima event, it would seem incumbent upon them to consider very carefully any and all decisions regarding nuclear energy. Even the USA, far across the same ocean, is quite vulnerable to such misguided past decisions, as Japan is vulnerable to bad US nuclear policy. Of course, there is poetic justice in this since much of the nuclear industry technology and infrastructure in Japan was sold and/or built by American companies.

Nevertheless, the Japanese were staring at a veritable sea of red flags at the very inception of the nation's nuclear energy policy. Having two cities destroyed, and upwards of 300,000 citizens killed by two atomic bombs, should have given them reason to pause. Being geographically situated in such an extraordinarily active seismic zone provided another opportunity to step back before the green light was given to nuclear power as energy policy.

Particularly in view of the awesome devastation which occurred during the Great Hanshin earthquake in Kobe in 1995, Japan had an excellent vantage point from which to observe the vulnerability of all its infrastructure. At that very moment in time, it was clear that the current building codes (and site locations) throughout the nuclear energy industry were wholly inadequate. Likewise, that the state of the art of nuclear technology was profoundly insufficient to withstand such earth-shaking assaults intact.

Furthermore, Japan has been way ahead of the curve where it concerns the acceleration of technospheric breakdown. Downtown Tokyo has given them many opportunities to observe and assess the wear and tear that is sustained each and every day by infrastructure and technology in an urban environment.

Every time Tokyo is hit by a typhoon or unsettled by an earthquake, the entire city experiences damage, much of which is unspoken or unwritten. The amount of underground destruction to water mains, subways, fiberoptic internet cables, telecommunication wiring, and other infrastructure in the wake of a severe earthquake can be enormous. Some of the more hidden damage can go undetected for months, or even years. Eventually, after the ongoing technospheric breakdowns continue to whittle away at the integrity of the overall system or grid, it only takes a small trigger event to precipitate a system-wide failure … which can then cause a domino effect resulting in a regional collapse … or even a national event. As 'advanced' as Japan has become in the area of information technology, it ought to have become clear how things can break down very quickly because of the interpenetration of so many different realms essential to keeping a nuclear power plant running properly.

Now that may be 'okay' or quite resolvable with the array of problems that routinely present in the wake of a typhoon or earthquake, but it isn't acceptable where it concerns nuclear mishaps, as the world witnessed in Fukushima. We now live in a new world. With events such as Hurricane Katrina, the BP Gulf Oil Spill, the Fukushima Nuclear Disaster, Superstorm Sandy, 500 and 1000 year floods occurring yearly, and many other catastrophic events both natural and manmade, a new paradigm must be considered. Not to do so at this very late date is to do so at our collective peril.

Preventative Measures Must be Taken in the Meantime, Especially by Japan.

There are substantial preventative measures related to nuclear power plant configuration which can be taken by Japan, as well as by other countries in similar positions, in order to avert potential calamities like Fukushima. However, these can only be implemented with a full understanding of the Earth changes which have been visiting every corner of the globe. Whereas Japan has been stuck in denial about their NEP, they must also understand that, as a multi-island nation located out in the Pacific Ocean on the *Ring of Fire*, they are inordinately exposed. Typhoons, earthquakes, tsunamis, volcanoes (Wikipedia, List...) will certainly always pose threats; therefore, why not take the necessary preventative measures that can be taken to minimize the worst possible outcomes.

In the case of Japan's nuclear power plants, the best preventive option would be a systematic closing down of each and every plant on the four main islands of Japan. This could occur with the phasing in of other energy sources which would help meet national energy requirements. If this isn't happening already, it should have started on March 11, 2011.

This solution is being offered as the primary one of all possible solutions where it concerns Japan's Nuclear Energy Industry. The thesis should be transparent by now: When the nuclear energy paradigm collides with Earth changes and technospheric breakdown, there is only one answer. An unequivocal case has been made in this chapter for the deliberate closing down of the current nuclear power industry throughout Japan. There is no more wiggle room. And, by the way, wherever earthquakes occur in such numbers and severity, wiggle room is what it all comes down to.

As Technospheric Breakdown Accelerates, Nuclear Power Generation Mishaps will Increase and Intensify

There is really no way around this eventuality. As all the nuclear power plants age, they will succumb to the micro-stresses which inevitably occur in such an ever-deteriorating environment. Most people are unaware of the true depth and breadth of technospheric breakdown since it is a concept rarely taken up by academia or the media. The following excerpts provide a wider perspective of this unavoidable byproduct of the Industrial Revolution.

> Technospheric breakdown is something that occurs everywhere around the globe, 24/7, without interruption, and with tremendous repercussions. Let's start with anything that has been manufactured in the factories of the modern world or built on the surface of the Earth. Simply put, everything is in the constant state of breaking down, degeneration, deterioration.

246

What does this really mean when we say that every bridge is slowly breaking down, every road is in greater disrepair with each passing day, every reservoir is gradually degrading, every office building, every factory, every school, every home, etc. most of which adhered to very low building standards in the first place?

What does it mean when the infrastructure for every sewer system, municipal water division, electrical grid, airport, railway station, etc. is in a slow but sure process of degrading and breaking down. So, unfortunately, is every nuclear power plant across the planet. (Cosmic Convergence)

What makes this ongoing process of physical degradation so insidious is that it almost always occurs subliminally. Through a gathering array of various forces throughout post-modern civilization, there does exist a sort of conspiracy of circumstances which has greatly magnified the effects of technospheric breakdown. The completed marriage between the industrial base of the Western powers and the financial class throughout the world has guaranteed that this slow motion collapse will continue unabated. How so?

Because so many corporate decisions are made according to their impact on the bottom line, many inferior nuclear power plants have been constructed around the globe. Likewise, because the mega-banks and investment houses are now dictating to a financially-strapped Nuclear Energy Industry, substandard nuclear reactors have been designed, engineered and continue to be put into operation across the planet. One only has to take a close look at the websites dedicated to decommissioned nuclear reactors or cold shutdowns or partially closed nuclear power plants or emergency actions taken at various nuclear power generation sites to grasp just how precarious a position the entire industry is currently in.

Unknown to even many of the nuclear engineers who address these issues 'in the office', or who fix the cascade of problems at nuke plants themselves, is the notion of slow motion, subclinical, pernicious technospheric breakdown. It often manifests in ways where cause and effect cannot be easily established because of some of the unseen forces produced by atomic fission. With that said, it should be noted that a chapter could easily be dedicated to this particular issue alone, so significant is it to the future of nuclear power generation

Then there is the Problem of Nuclear Wastes and Natural Rights, Yes?

No one has articulated this point better than Albert Bates in his definitive essay entitled The Karma of Kerma: Nuclear Wastes and Natural Rights (Bates, A.K., 1988)

This extremely lucid and illuminating, sober and sane treatment of the greatest ongoing environmental disaster of our times lays bare the most basic legal and human rights issues which converge around the production, treatment and storage of nuclear wastes. Were the governments of the world to read and take to heart its simple and straightforward thesis, the current incarnation of nuclear energy production would have been abandoned years ago:

> The disposal of radioactive substances in a manner that anticipates their eventual partial release into the human environment imposes a health burden upon future generations that cannot be justified by any moral or legal rationale. Like an irresistible force meeting an immovable object, the concept of the greater good for the many in the present generation runs against the concept of the inalienable rights of each individual in future eras. At present, in matters involving nuclear power, our governmental agencies have taken the side of the irresistible force. But when federal agencies venture to tread beyond of the scope of the foundation principles with which the federal government was fashioned, they endanger more than human lives. At risk in the nuclear waste debate are long-held concepts of ordered liberty. (Bates A. K., 1988)

Fukushima has illustrated exactly why this elegantly stated legal concept of human rights and moral imperative is so pertinent to the public discourse. When massive amounts of radioactive wastewater are dumped into the Pacific Ocean, not only human life will be adversely affected. Marine life has been negatively impacted in ways that will take decades to observe and comprehend. The outright destruction of the environment in and around Fukushima and the Pacific Ocean must also be considered in any meaningful assessment of collateral damage.

Perhaps even more than Chernobyl, Fukushima has allowed the global community to view the whole event through the lens of legal responsibility and ethical outcomes so that new international standards can be written and implemented regarding nuclear waste conveyance and disposal. If nothing else, this discussion has raised awareness about the most nagging issue concerning the NEP. Whereas the human rights aspect confers the legal right to not be contaminated by nuclear radiation has barely been addressed by those responsible for it consequences, it now enjoys a prominent place throughout the worldwide debate.

Fukushima: Wakeup Call of the Millennium

Accidents and mishaps, manmade and natural disasters happen. Things are fixed fairly quickly in this postmodern age, and life goes on. Whether these events occur in a full-blown war zone or in the wake of a hurricane, the affected population usually does everything it can to rebuild and move on.

However, when these events take place in or near nuclear power plants, life doesn't just go on. It often stops. Depending on the circumstances and seriousness of a nuclear event, sometimes life stops in that area for a long time.

Our civilization has now been given three unmistakable wakeup calls since the advent of the nuclear power generation era. First there was Three Mile Island in Pennsylvania, then there was Chernobyl in the Ukraine, and lastly the world is still reeling from the specter of possibilities which are presented by Fukushima.

Surely it is not by chance that these three flagrant examples of nuclear *Perfect Storms* occurred around the globe affecting major nations and populations centers. Each of these disasters has served to wake up whole swaths of humanity to the dangers and risks which are associated with the current *Nuclear Energy Paradigm*. To ignore, or deny, or refute the obvious lessons which all three nuclear catastrophes have given to humankind would be folly of the highest order.

The global impact of Fukushima, which has disseminated radionuclides (radioactive contaminants) by air and by way of the largest of the seven seas, stands as dramatic testimony to all that can go wrong -- seriously wrong -- with the current nuclear energy business model and method of power generation. Can it get any worse than Fukushima? That we are compelled to even ask this question speaks volumes about the true state of the affairs on that 25 square mile patch of land and contiguous sea which surround the Fukushima Daiichi nuclear disaster site.

Given this inescapable testament of nuclear folly, it is now incumbent upon the community of nations to rally around the obvious necessity of terminating the current form of the *Nuclear Energy Paradigm*. Why? Because when a "China Syndrome" occurs anywhere in the world, it will inevitably affect the entire planet. In other words, an INES Level 7 (Wikipedia, International Nuclear Event Scale) nuclear catastrophe does not respect borders. Nor does it discriminate between the young and old, healthy and sick, or those who live close to ground zero from those who live far away.

Therefore, any nation that chooses to set up a nuclear energy-producing operation from this point forward has an inviolable responsibility to its neighbors, as it does to the rest of the world. Likewise, those nations have a moral obligation to proceed in a manner that guarantees its neighbors will not be exposed to the consequences of its nuclear accidents, even when they are caused by duel natural disaster events as we saw at Fukushima.

Just as Europe was contaminated with radiation from Chernobyl (Yablokov, A.V., 2009), and North America has been contaminated from Fukushima, it is understood that once a nuclear catastrophe spirals out of control, the genie of radioactive contamination cannot be put back in the bottle. The entire Pacific Rim, in fact, has varying degrees of exposure to the radioactive waste water being conveyed by the ocean from Fukushima, as does the Western Hemisphere to seaborne radioactive isotopes like Cesium-137 and airborne isotopes such as Iodine-131(Center for Marine and Environmental Radiation).

Consequently, Japan is responsible for the damage wrought to the largest ocean on Earth. Have they acknowledged this? Have they approached the nations

249

both near and far which have been affected by their cavalier and irresponsible approach to siting reactors up and down their seismic shorelines? Has the United Nations even addressed this extremely important issue known as national accountability? Or territorial sovereignty?

Conclusion

It doesn't get very much more weighty than the 'fallout from Fukushima'. All of the affected nations have been curiously silent on this issue. It is almost as though a conspiracy of silence has descended upon the concerned countries because of how unpredictable and intractable the nuclear containment problems have been at the Daiichi plant.

At the end of the day the current race of humanity will look back on the Fukushima Nuclear Disaster as the defining moment for both the industry and the underlying paradigm. It they haven't already, the various stakeholders will be forced to re-evaluate the integrity of their nuclear enterprises around the globe. Hopefully, they will begin to take aggressive preemptive measures to address whatever needs to be addressed at every nuclear site still in operation.

If a decisive response is not formulated and implemented on a global scale, in light of the hard lessons learned from Fukushima, the current planetary civilization will be compelled to face up to these fatal flaws in most unpleasant ways, which will continue to manifest with each major Earth change. In a similar way, the inherent defects of the NEP will only be accentuated as technospheric breakdown accelerates. The profound and fundamental shortcomings which pervade the entire nuclear energy industry can no longer be hidden or ignored.

After all, it was the dangerous combination of willful blindness and feigned ignorance which got the world into this position in the first place.

> Does anyone in their right mind believe that nuclear power plants can ever be designed, engineered or constructed to withstand 9.0 earthquakes followed by 15 meter high tsunamis? Sorry if we offend, but such a display of so deadly a combination of ignorance and arrogance must represent the very height of hubris. Particularly in view of the inevitable consequences which have manifested at Fukushima, how is it that so few saw this pre-ordained and disastrous outcome, except by willful blindness? (State of the Nation)

Japan should never have sited 55 nuclear reactors (plus 12 others) on its coastlines. Which begs the question: Why did they?

The following observation has been made by many experts and laypeople alike since March 11 of 2011.

> Quite purposefully, no one ever stopped to consider the obvious and far-reaching ramifications of constructing 55 nuclear

reactors on the most seismically active piece of property on Planet Earth! And, that doesn't count another 12 reactors in various stages of planning or development. (State of the Nation)

Graphic Links

[1] Japan NPP map
http://cosmicconvergence.org/wpcontent/uploads/2014/07/japan.jpg

[2] 45-foot high Tsunami
https://www.youtube.com/watch?v=G85Jy5ff7Q&index=1&list=PLkCTP872qR8c_elZBQ_F-xZJOZEFCdyd9

[3] Fort Calhoun Nuclear Power Station during the floods of June, 2011 (1)
http://cosmicconvergence.org/wp-content/uploads/2014/07/fort-calhoun-nuclear-station.top_.jpg

[4] Fort Calhoun Nuclear Power Station during the floods of June, 2011 (2)
http://cosmicconvergence.org/wp-content/uploads/2014/07/1280px-Calhoun-Corp_of_Eng._6-16-11A_266.jpg

[5] Devastation at TEPCO's Fukushima Daiichi Nuclear Power Station (1)
http://cosmicconvergence.org/wp-content/uploads/2014/07/pict5.jpg

[6] Devastation at TEPCO's Fukushima Daiichi Nuclear Power Station (2)
http://cosmicconvergence.org/wp-content/uploads/2014/07/pict12.jpg

[7] Devastation at TEPCO's Fukushima Daiichi Nuclear Power Station (3)
http://cosmicconvergence.org/wp-content/uploads/2014/07/pict13.jpg

References

Associated Press. Magnitude 6.8 earthquake off Japan coast triggers tsunami. July 11, 2014. http://www.wjla.com/articles/2014/07/magnitude-6-8-earthquake-off-japan-coast-tsunami-alert-issued-105009.html

Bates, A. K., Karma of Kerma. Journal of Environmental Law and Litigation. Univ. of Oregon School of Law, Vol 9, Page 3, February, 1988. http://www.thefarm.org/lifestyle/albertbates/akbp5.html

Blodget, H. The US Is Having A Catastrophic Nuclear Emergency In Nebraska And The Obama Administration Is Covering It Up. Russia, Business Insider, June 19, 2011. http://www.businessinsider.com/nebraska-nuclear-meltdown-2011-6

Center for Marine and Environmental Radiation, Woods Hole Oceanographic Institution. How Radioactive is Our Ocean? http://www.ourradioactiveocean.org/

Cosmic Convergence. Technospheric Breakdown: Ongoing, Ubiquitous and Unstoppable. January 2, 2012. http://cosmicconvergence.org/?p=3717So%2c%20unfortunately%2c%20is%20every%20nuclear%20reactor%20across%20the%20planet.%22

Davies, A. NYC Has Some Big Ideas To Save Its Subways From The Next Superstorm, Business Insider, June 13, 2013. http://www.businessinsider.com/nycs-subway-isnt-ready-for-another-sandy-2013-6

Forbes Magazine, February 1986. In Nuclear Reader, Chapter Two, Nuclear Power is Bad Business. http://www.nuclearreader.info/chapter2.html

Funk, J. and Nelson Lampe. Fort Calhoun Nuclear Plant: Flood Seeps Into Turbine Building At Nebraska Nuke Station, Huffington Post, 06/27/11. http://www.huffingtonpost.com/2011/06/27/fort-calhoun-nuclear-flood-nebraska-plant_n_885067.html

Gunter, P. Hazards of Boiling Water Reactors in the United States - Nuclear Information and Resource Service (NIRS), March, 1996, updated by Michael Mariotte, NIRS, March 2011 http://www.nirs.org/factsheets/bwrfact.htm

Hargreaves, S. Flooded Nebraska nuclear plant raises broader disaster fears, CNN. June 28, 2011. http://money.cnn.com/2011/06/28/news/economy/nebraska_nuclear_plant/

Kageyama, K.Typhoon Neoguri Slams Into Okinawa, Generating 46-Foot Waves, Huffington Post. 07/07/2014. http://www.huffingtonpost.com/2014/07/07/typhoon-neoguri_n_5562427.html

State of the Nation. The Pacific Ocean Is Dying: A Special Report On the Fukushima Nuclear Catastrophe, October 29, 2013. http://stateofthenation2012.com/?p=2289

Sulzberger, A. G. and Matthew L. Wald, Flooding Brings Worries Over Two Nuclear Plants. The New York Times, June 20, 2011. http://www.nytimes.com/2011/06/21/us/21flood.html?_r=1&

The Doc. Fort Calhoun Nuke Plant Flood Wall Collapse The Result Of Accidental Puncture, SilverDoctors, June 26, 2011. http://www.silverdoctors.com/fort-calhoun-nuke-plant-flood-wall-collapse-the-result-of-accidental-puncture/

Tokaimura Nuclear Accident http://en.wikipedia.org/wiki/Tokaimura_nuclear_accident

U.S. Congress, Office of Technology Assessment, *Aging Nuclear Power Plants: Managing Plant Life and Decommissioning, OTA-E-575* (Washington, DC: U.S. Government Printing Office, September 1993). (Chapter 2, Safety of Ageing Nuclear Plants, p.38) Please see the following URL for the material mentioned here: http://cosmicconvergence.org/wp-content/uploads/2014/07/Screen-Shot-2014-07-15-at-6.31.48-AM.png

Wald, M.L. Nuclear Plants, Old and Uncompetitive, Are Closing Earlier Than Expected. June 14, 2013, The New York Times. http://www.nytimes.com/2013/06/15/business/energy-environment/aging-nuclear-plants-are-closing-but-for-economic-reasons.html?_r=0

Washington's Blog. Fukushima: General Electric Knew Its Nuclear Reactor Design
Was Unsafe ... So Why Isn't GE Getting Any Heat for Fukushima? Global
Research, December 12, 2013. http://www.globalresearch.ca/fukushima-
general-electric-knew-its-nuclear-reactor-design-was-unsafe-so-why-isnt-ge-
getting-any-heat-for-fukushima/5361300

Wikipedia, International Nuclear Event Scale.
http://en.wikipedia.org/wiki/International_Nuclear_Event_Scale

Wikipedia, List of volcanoes in Japan.
http://en.wikipedia.org/wiki/List_of_volcanoes_in_Japan

World Awash in Environmental Armageddon, Japan: A Nation Consigned To
Nuclear Armageddon, An Open Letter to the People of Japan. March 17, 2011.
http://environmentalarmageddon.wordpress.com/2011/03/17/japan-a-nation-
consigned-to-nuclear-armageddon/

Yablokov, A.V., Vassily B. Nesterenko. Alexey V. Nesterenko. 2009. Chernobyl
Consequences of the Catastrophe for People and the Environment, New York
Academy of Sciences.
http://environmentalarmageddon.files.wordpress.com/2011/03/chernobyl-
sm.pdf

CONCLUSION

Fukushima significantly increased the world's "background" level of radionuclides: "During the passage of contaminated air masses from Fukushima, airborne 137Cs levels were globally enhanced by 2 to 3 orders of magnitude" (cited in Masson et al., 2013; Masson et al., 2011) while the "129I/127I ratio in Fukushima precipitation in March 2011 immediately after the Fukushima accident was more than 3 orders of magnitude higher than the background level of this region" (Xu et al., 2013, p. 10851). People living in heavily contaminated areas in Japan are expected to face up to a 4 Sievert lifetime dose from external exposure alone (see Ioannidou et al., 2014, p. 858). Contamination of the ocean, fresh water aquifer, and atmosphere are ongoing.

The nuclear village denies the special risks posed by nuclear energy by failing to recognize full hazards and costs of nuclear contamination. Paul Langley, Chris Busby and Harvey Wasserman demonstrate in this collection that public denial has been the modus operandi since atomic bombs were dropped on Hiroshima and Nagasaki. Denial of risks to human health is so well-institutionalized that the contemporary response by government to radiation emergencies, such as the one posed by Fukushima, is to raise permissible exposure standards, a practice which occurred in Japan and is being promoted in the US. Environmental externalities from decades of contamination by radionuclides are simply absorbed by the eco-system, further degrading its carrying capacity. Human health impacts include increased disease and reproductive problems across generations, incurring incalculable costs and suffering. Destruction of earth's collective genetic heritage and carrying capacity are the inevitable outcome of the dance of destruction wrought by the nuclear paradigm, argues Majia Nadesan. Destruction is the teleology of nuclear bombs and the unintended consequence of the entire nuclear energy cycle.

Nuclear power is politically disenfranchising and is incommensurable with free market capitalism, as documented in this book by Andrew McKillop and Christian T. Lystbaek. Nuclear power disenfranchises politically by destroying personal property and threatening human health through routine and catastrophic emissions. The global nuclear-military-industrial complex denies its externalities with flawed scientific dose models and through often-repeated practices of obfuscating, marginalizing and trivializing contamination. Denial of externalities has allowed nuclear-power industries to *greenwash* operations, excluding from calculations of environmental impact the entire fuel cycle, ranging from uranium mining, to de-commissioning and long-term storage of radioactive waste. Governments have facilitated denial by underwriting enormous "decommissioning" and accident mitigation costs. Full accounting of the range and severity of environmental and health impacts would belie claims to market fitness and sustainability. As Lystbaek argues, "it is not the market that fails to account for the costs of nuclear power, but the cost of nuclear power fails to qualify for a market."

Nuclear power prevails because it is part of a military-industrial complex that spans nations. The nuclear machine wants to replicate. As discussed by Adam Broinowski, supposedly non-nuclear weapons states such as Japan become complicit in the military aspects of nuclear power through their involvement in the transnational nuclear industry. This is due to the interdependency imposed by both the long-term economic and trade commitments required for nuclear technologies and by the security agreements which bind such nations to nuclear deterrence regimes. Newly nuclear nations implicitly align with the states that bequeath their membership in the global nuclear club. Nuclear energy was always recognized as posing risks for proliferation of nuclear weapons capabilities. However, the will to power prevailed over rational risk assessments. Today, the global nuclear cabal of warring nation states and their lapdog nuclear industries threaten us all with extinction. Nuclear cabals imagine themselves after gods because they falsely believe they can control the atom. Fukushima and every other nuclear disaster have demonstrated their failure, their false hubris, a story that has been compellingly narrated in this volume and other locales by Harvey Wasserman.

There are two paths forward. One path is to repeat the cycle of death and destruction represented by William Banzai7 in his "Hiroshima-Fukushima" cover-art composition. Japan's rising militarism and deteriorating nuclear power infrastructure both promise future radiation catastrophes, as described by The Fukushima Five in their analysis of technospheric breakdown. The alternative path requires a gestalt-switch prioritizing truly *sustainable* forms of energy production and social organization. The path toward sustainable energy and social organization requires full transparency of costs, including potential externalities, and truly democratic decision-processes. Tony Boys and Richard Wilcox explain how local politics can be shaped covertly by national, and perhaps global, nuclear forces, complicating citizen activism. Yet, despite entrenched legacy industries and widespread propaganda, many citizens in Japan and elsewhere identify with localized and alternative sources of energy, as documented by Boys and Wilcox. Andrew McKillop helps map strategic change by charting failures and successes in energy transition in Europe and Japan. The will and capacity for change exist, but must be nourished by greater government transparency and accountability, corporate social responsibility, public education, and individual willingness to confront power in order to ensure our very survival.

References

Ioannidou, A., Manolopoulou, E. M., Stoulos, S., Vagena, E., Papastefanou, C., Bonardi, M. L., Gin, L., Manenti, S., & F. Groppi (2014). Radionuclides from Fukushima accident in Thessaloniki, Greece (40°N) and Milano, Italy (45°N). *Journal of Radioanalytical and Nuclear Chemistry*, 299(1), 855-860.

Masson, O., Ringer, W., Malá, H., Rulik, P., Dlugosz-Lisiecka, M., Eleftheriadis, K., Meisenberg, O., De Vismes-Ott, A. & F. Gensdarmes (2013). Size distributions of airborne radionuclides from the Fukushima nuclear accident at several places in Europe. *Environmental Science and Technology*, 47 (19), 10995–11003.

Masson, O. et al., (2011). Tracking of airborne radionuclides from the damaged Fukushima Dai-Ichi Nuclear Reactors by European networks. *Environmental Science and Technology*, 45 (18), 7670−7677.

Xu, S., Stewart P. H. T. Freeman, S. P. H. T., Hou, X., Watanabe, A., Yamaguchi, K., & L. Zhang (2013). Iodine Isotopes in Precipitation: Temporal Responses to 129I Emissions from the Fukushima Nuclear Accident. *Environmental Science and Technology*, 47, 10851-1085

INDEX

US National Academy of Sciences, 18, 80
USS Ronald Reagan, 166, 170
Utsunomiya, K. 57-61, 63-65, 71

V

vehicle air filters, 90

W

Watanabe, Y. 59
weapons fallout, 77, 81, 91, 94
West Coast, 94, 171
Westinghouse, 10, 14, 27, 132, 151
wind power, 200, 203, 216
Windscale, 7-9, 43, 134
Woods Hole Oceanographic Institution, 135, 251

World Health Organization (WHO), 44, 75, 80, 81, 89, 165

X

X-rays, 75, 76, 144

Y

Yablokov, A.V. 81, 99, 113, 125, 170, 249, 253
yakuza, 64
Yanagida, K. 116, 117
Yanagihara, T. 89
Yaroshinskaya, A. 77
Yomiuri Shinbun, 36, 38, 45
Yonekura, M. 57
Your Party, 59

www.ingramcontent.com/pod-product-compliance
Lightning Source LLC
Chambersburg PA
CBHW031829170526
45157CB00001B/236